SHOW WHAT

CSAP

FOR GRADE 8

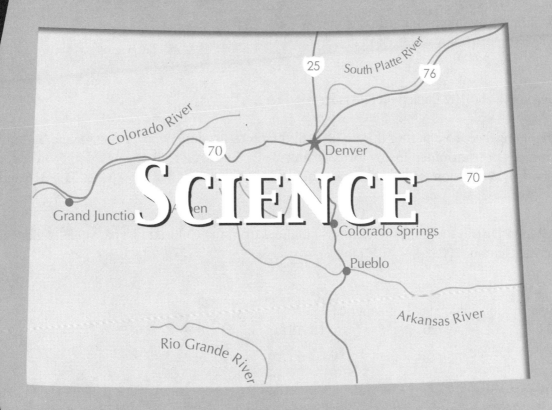

grade
8

**PREPARATION FOR THE
COLORADO STUDENT
ASSESSMENT PROGRAM**

Published by:

Show What You Know® Publishing
A Division of Englefield & Associates, Inc.
P.O. Box 341348
Columbus, OH 43234-1348
Phone: 614-764-1211
www.showwhatyouknowpublishing.com
www.passthecsap.com

Copyright © 2007 by Englefield & Associates, Inc.

All rights reserved. No part of this book, including interior design, cover design, and icons, may be reproduced or transmitted in any form, by any means (electronic, photocopying, recording, or otherwise). Permission is granted to reproduce the Correlation Charts on pages 314–316 and 376–378.

CSAP Item Distribution information was obtained from the Colorado Department of Education Web site, October 2007.

Printed in the United States of America
09 08 07   20 19 18 17 16 15 14 13 12 11 10 9 8 7 6 5 4 3 2 1

ISBN: 1-59230-250-5

Limit of Liability/Disclaimer of Warranty: The authors and publishers have used their best efforts in preparing this book. Englefield & Associates, Inc., and the authors make no representations or warranties with respect to the contents of this book and specifically disclaim any implied warranties and shall in no event be liable for any loss of any kind including but not limited to special, incidental, consequential, or other damages.

Acknowledgements

Show What You Know® Publishing acknowledges the following for their efforts in making this assessment material available for Colorado students, parents, and teachers.

Cindi Englefield, President/Publisher
Eloise Boehm-Sasala, Vice President/Managing Editor
Christine Filippetti, Production Editor
Jill Borish, Production Editor
Charles V. Jackson, Mathematics Editor
Trisha Barker, Assistant Editor
Angela Gorter, Assistant Editor
Jennifer Harney, Editor/Illustrator

About the Contributors

The content of this book was written BY teachers FOR teachers and students and was designed specifically for the Colorado Student Assessment Program (CSAP) for Grade 8 Science. Contributions to the Science section of this book also were made by the educational publishing staff at Show What You Know® Publishing. Dr. Jolie S. Brams, a clinical child and family psychologist, is the contributing author of the Test Anxiety and Test-Taking Strategies chapters of this book. Without the contributions of these people, this book would not be possible.

Table of Contents

Introduction v

Test Anxiety 1

Test-Taking Strategies for Science 11

Science 23
    Introduction 23
    Understanding Assessment Objectives 24
    Colorado's Science Standards 25
    About the Science CSAP 37
    Glossary of Science Terms 40
    Science Practice Tutorial
        Directions for Science Practice Tutorial 55
        Sample Science CSAP Questions 56
        Science Practice Tutorial 58
    Science Assessment One
        Directions 253
        Assessment One—Session One 254
        Assessment One—Session Two 267
        Assessment One—Session Three 282
        Assessment One—Answer Key 299
        Assessment One—Correlation Charts 314
    Science Assessment Two
        Directions 317
        Assessment Two—Session One 318
        Assessment Two—Session Two 332
        Assessment Two—Session Three 347
        Assessment Two—Answer Key 361
        Assessment Two—Correlation Charts 376
    CSAP Science Standards Checklist 379

Introduction

Dear Student:

This *Show What You Know® on the CSAP for Grade 8 Science, Student Self-Study Workbook* was created to give you practice in preparation for the Colorado Student Assessment Program (CSAP) in Science.

The first chapter in this workbook—Test Anxiety—was written especially for eighth-grade students. Test Anxiety offers advice on how to overcome nervous feelings you may have about tests.

The Test-Taking Strategies chapter includes helpful tips on how to answer questions correctly so you can succeed on the Science section of the CSAP.

In the Science chapter of this book, you will find the following:
- A Science Glossary to review Science terms.
- Scoring Guides for multiple-choice and constructed-response questions.
- A Science Practice Tutorial with multiple-choice and constructed-response questions. An analysis for each Practice Tutorial question is given to help you identify the correct answer.
- Two full-length Science Assessments follow the Practice Tutorial section for additional Science practice. An Answer Key will help you check to see if you answered the Assessment questions correctly.

This Student Self-Study Workbook also include Correlation Charts for Science. These charts can be used to identify individual areas of needed improvement.

This book will help you become familiar with the look and feel of the Science CSAP Assessments and will provide you with a chance to practice your test-taking skills to show what you know.

Good luck on the CSAP!

Test Anxiety

What Is Test Anxiety?

Test anxiety is a fancy term for feelings of worry and uneasiness that students feel before or during a test. Almost everyone experiences some anxiety at one time or another. Experiencing feelings of anxiety before any challenge is a normal part of life. However, when worrying about tests becomes so intense it interferes with test taking, or if worrying causes students mental or physical distress, this is called test anxiety.

Test Anxiety CSAP Science for Grade 8

What Are the Signs of Test Anxiety?

Test anxiety is much more than feeling nervous. In fact, students will notice test anxiety in four different areas: thoughts, feelings, behaviors, and physical symptoms. No wonder test anxiety gets in the way of students doing or feeling well.

1. Thoughts

Students with test anxiety usually feel overwhelmed with negative thoughts about tests and about themselves. These thoughts interfere with the ability to study and to take tests. Usually, these bothersome thoughts fall into three categories:

- **Worrying about performance**—A student who worries may have thoughts such as, "I don't know anything. What's the matter with me? I should have studied more. My mind is blank; now I'll never get the answer. I can't remember a thing; this always happens to me. I knew this stuff yesterday and now I can't do anything."

- **Comparing oneself to others**—A student who compares performance might say, "I know everyone does better than I do. I'm going to be the last one to finish this. Why does everything come easier for everyone else? I don't know why I have to be different than others."

- **Thinking about possible negative consequences**—A student with negative thoughts would think, "If I don't do well on this test, my classmates will make fun of me. If I don't do well on the Science CSAP, my guidance counselor will think less of me. I won't be able to go to my favorite college. My parents are going to be angry."

Many of us worry or have negative thoughts from time to time. However, students with test anxiety have no escape and feel this worry whenever they study or take tests.

2. Feelings

In addition to having negative thoughts, students with test anxiety are buried by negative feelings. Students with test anxiety often feel:

- **Nervous and anxious**—Students feel jittery or jumpy. Anxious feelings may not only disrupt test taking but may interfere with a student's life in other ways. Small obstacles, such as misplacing a book, forgetting an assignment, or having a mild disagreement with a friend, may easily upset students. They may become preoccupied with fear, may have poor self-esteem, and may feel that the weight of the world is on their shoulders. They seem to be waiting for "the next bad thing to happen."

- **Confused and unfocused**—Students with test anxiety have their minds in hundreds of anxious places. They find it difficult to focus on their work, which makes studying for tests even harder. Students with test anxiety also have difficulty concentrating in other areas. When they should be listening in class, their minds worry about poor grades and test scores. They jump to conclusions about the difficulty of an upcoming test. They find themselves fidgeting. They constantly interrupt themselves while studying, or they forget how to complete simple assignments. Anxiety can interfere with a student's ability to focus, study, and learn.

- **Angry and resentful**—Test anxiety can lead to irritable and angry feelings. Anxious students are defensive when communicating with others. They become overwhelmed by negative thoughts and feel they are not good enough. Test anxiety also makes students feel "trapped" and as though they have no escape from school or tests. Students who feel there is no way out may get angry; they may resent the situation. They feel jealous of people they believe find school easier. They are angry at the demands placed on them. The more angry and resentful students become, the more isolated and alone they feel. This only leads to further anxiety and increased difficulties in their lives.

- **Depressed**—Anxiety and stress can lead to depression. Depression sometimes comes from "learned helplessness." When people feel they can never reach a goal and that they are never good enough to do anything, they tend to give up. Students who are overly anxious may get depressed. They lose interest in activities because they feel preoccupied with their worries about tests and school. It might seem as though they have no time or energy for anything. Some students with test anxiety give up on themselves completely, believing if they cannot do well in school (even though this may not be true), then why bother with anything?

Not all students with test anxiety have these feelings. However, if you or anyone you know seem to be overwhelmed by school, feel negative most of the time, or feel hopeless about school work (test taking included), you should look to a responsible adult for some guidance.

3. Behavior

Students with test anxiety often engage in behavior that gets in the way of doing well. When students have negative thoughts and feelings about tests, they participate in counterproductive behavior. In other words, they do things that are the opposite of helpful. Some students avoid tests altogether. Other students give up. Other students become rude and sarcastic, making fun of school, tests, and anything to do with learning. This is their way of saying, "We don't care." The truth is, they feel anxious and frustrated. Their negative behaviors are the result of thoughts and feelings that get in the way of their studying and test taking.

4. Physical Symptoms

All types of anxiety, especially test anxiety, can lead to very uncomfortable physical symptoms. Thoughts control the ways in which our bodies react, and this is certainly true when students are worried about test taking. Students with test anxiety may experience the following physical symptoms at one time or another:

- sweaty palms
- stomach pains
- "butterflies" in the stomach
- difficulty breathing
- feelings of dizziness or nausea
- headaches
- dry mouth
- difficulty sleeping, especially before a test
- decrease or increase in appetite

Test anxiety can cause real physical symptoms. These symptoms are not made up or only in your head. The mind and body work together when stressed, and students can develop uncomfortable physical problems when they are anxious, especially when facing a major challenge like the CSAP.

The Test Anxiety Cycle

Have you ever heard the statement "one thing leads to another"? Oftentimes, when we think of that statement, we imagine Event A causes Event B, which leads to Event C. For example, being rude to your younger brother leads to an argument, which leads to upset parents, which leads to some type of punishment, like grounding. Unfortunately, in life, especially regarding test anxiety, the situation is more complicated. Although one thing does lead to another, each part of the equation makes everything else worse, and the cycle just goes on and on.

Let's think back again to teasing your younger brother. You tease your younger brother and he gets upset. The two of you start arguing and your parents become involved. Eventually, you get grounded. Sounds simple? It might get more complicated. When you are grounded, you might become irritable and angry. This causes you to tease your little brother more. He tells your parents, and you are punished again. This makes you even angrier, and now you don't just tease your little brother, you hide his favorite toy. This really angers your parents who now do not let you go to a school activity. That upsets you so much you leave the house and create trouble for yourself. One thing feeds the next. Well, the same pattern happens in the test anxiety cycle.

Look at the following diagram.

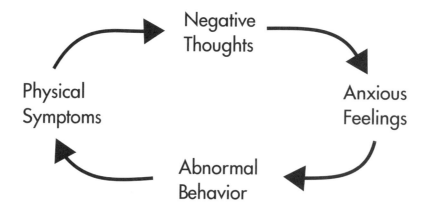

As you can see, the four parts of this diagram include the thoughts, feelings, behavior, and physical symptoms discussed earlier. When a student has test anxiety, each area makes the others worse. The cycle continues on and on. Here's an example:

> Let's start off with some symptoms of negative thinking. Some students might say to themselves, "I'll never be able to pass the Science CSAP Assessment!" This leads to feelings of frustration and anxiety. Because the student has these negative thoughts and feelings, his or her behavior changes. The student avoids tests and studying because they are nerve racking. Physical symptoms develop, such as the heart racing or the palms sweating. Negative thoughts then continue, "Look how terrible I feel; this is more proof I can't do well." The student becomes more irritable, even depressed. This affects behavioral symptoms again, making the student either more likely to avoid tests or perhaps not care about tests. The cycle goes on and on and on.

Is Test Anxiety Ever Good?

Believe it or not, a little worrying can go a long way! Too much test anxiety gets in the way of doing one's best, but students with no anxiety may also do poorly. Studies have shown that an average amount of anxiety can help people focus on tasks and challenges. This focus helps them use their skills when needed. Think about a sporting event. Whether a coach is preparing an individual ice skater for a competition or is preparing the football team for the Friday night game, getting each athlete "psyched up" can lead to a successful performance. A coach or trainer does not want to overwhelm the athlete. However, the coach wants to sharpen the senses and encourage energetic feelings and positive motivation. Some schools have a team dinner the night before a competition. This dinner provides some pleasant entertainment, but it also focuses everyone on the responsibilities they will have the next day.

Consider the graph below. You can see that too little test anxiety does not result in good test scores. As students become more concerned about tests, they tend to do better. But wait! What happens when too much anxiety is put into the equation? At that point, student performance decreases remarkably. When anxiety reaches a peak, students become frustrated and flustered. Their minds tend to blank out, they develop physical symptoms, they cannot focus, and they also behave in ways that interfere with their performance on tests.

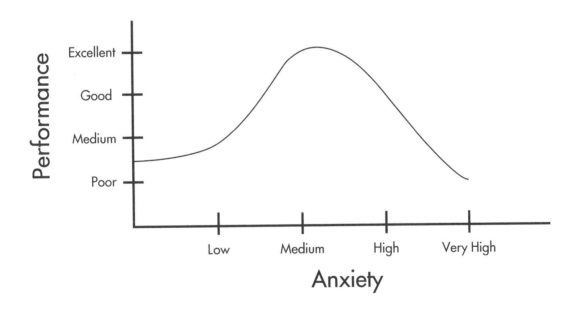

An important key to successful test taking is to get yourself in the right mood about taking a test. Looking at a test as a challenge and looking forward to meeting that challenge, regardless of the end result, is a positive and healthy attitude. You will feel excited, motivated, and maybe a little nervous but certainly ready to face the CSAP.

How Do I Tackle Test Anxiety?

Although test anxiety is an uncomfortable and frustrating feeling, the good news is you can win the battle over test anxiety! Conquering test anxiety will not be accomplished by luck or magic, but it can be done by students of all ages in a relatively short period of time. If you can learn to master test anxiety at this point in your life, you will be on the road to successfully facing many other challenges you will encounter.

1. Change the way you think

Whether you realize it or not, your thoughts—good and bad—influence your life. The way we think is related to how we feel about ourselves, how we get along with other people, and how well we do in school, especially when taking tests.

- **Positive thinking can block out negative thinking**—It is impossible to think two opposite thoughts at the same time. You may have one idea and then think about another, but one is always going to "win" over the other. When you practice positive thinking, you are replacing negative thoughts with positive ones. The more you are able to think positive thoughts, the less you will be troubled by negative ones.

- **The soda pop test**—It's just as easy to have positive thoughts as negative ones. Everyone has heard the saying, "There is more than one side to any story." Just as there are two opinions on any given subject, there is generally more than one way to look at almost every situation in life. Some ways are more helpful than others.

Think about a can of soda pop. Draw a line down the middle of a blank piece of paper. On one side, put the heading, "All the bad things about this can of soda pop." On the other side, put another heading, "All the good things about this can of soda pop." Now, write appropriate descriptions or comments under each heading. For example, you could write, "This can of soda pop is a lot smaller than a two-liter bottle," which is negative thinking. Or, you could write, "This can of soda pop is just the right size to stay cold and fizzy until I finish it." It's easy to look at the soda pop can and think bad thoughts. But you also are able to come up with many good things. If you spent all your time focusing on the negative aspects, you might believe the can of soda pop is bad. It is better to look at the positive side of things.

Part of successful test taking has to do with how you look at tests. With the can of soda pop, you could choose to think negatively, or you could have positive thoughts. The same holds true for tests. You can look at a test as a scary or miserable experience, or you can look at a test as just one of many challenges you will face in your life.

Counselors have known for years that people who are worried or anxious can become happier when thinking positive thoughts. Even when situations are scary, such as going to the dentist or having a medical test, "positive imagery" is very helpful. Positive imagery simply means focusing on good thoughts to replace anxious thoughts.

You can replace negative thoughts with positive ones through practice. Believe it or not, it really works!

- **Thoughts of success**—Thinking "I can do it" thoughts chases away ideas of failure. Times that you were successful, such as when you did well in a sports event or finished a challenging science experiment, are good things to think about. Telling yourself you have been successful in the past and can now master the Science CSAP will replace thoughts that might otherwise cause anxiety.

- **Relaxing thoughts**—Some people find that thinking calming or relaxing thoughts is helpful. Picturing a time in which you felt comfortable and happy can lessen your anxious feelings. Imagining a time when you visited the ocean, climbed a tree, or attended a concert can help you distract your mind from negative thoughts and focus on times that you were relaxed, happy, and felt positive.

- **All-or-nothing thinking**—Nothing is ever as simple as it seems. Sometimes we convince ourselves something is going to be "awful" or "wonderful," but it rarely turns out that way.

No test is "completely awful" or "completely perfect." Tests are going to have easy questions and hard questions, and you are going to have good test days and bad test days. The more you set up expectations that are all positive or negative, the more stressful the situation becomes. Accepting that nothing is totally good or bad, fun or boring, or easy or hard will reduce your anxiety and help you set reasonable expectations about tests. When you think about tests, try not to think about them as the road to academic success or a pit of failure. Instead, realize that all challenges have both good and bad elements, and you have to take everything in stride.

- **Making "should" statements**—Making "should" statements sets students up for failure. Sure, it is important to try your best, to study hard, and to make a reasonable effort on the CSAP; it may even be good to take an extra study session, try another practice test, or ask a teacher or tutor for advice and suggestions. It is also a good idea to use a book, such as this one, to help you do your best and show what you know. However, there is a big difference between doing your "reasonable best" and living your life with constant worries and put-downs. Students who constantly tell themselves "I should" and berate themselves for not having done everything possible only increase their levels of anxiety.

Go back to the test anxiety cycle. Suppose your thoughts are, "I should have stayed up an extra hour and studied" or "I should have reviewed the rock cycle." The more you think these thoughts, the more anxious you get. The more anxious you get, the worse you feel. Again, the cycle goes on and on.

One part of maturing is learning to balance your life. Life is happiest when you find a good balance between being a lazy do-nothing and being a perfectionist. While we all know laziness gets us nowhere, being a perfectionist may actually paralyze your future chances of success because you will eventually fear meeting any new challenges. Failure does not mean real failure; it just means being imperfect. Preventing perfectionism begins by saying "no" to unreasonable thoughts and "should" statements. "Should" statements place high demands on a student and only lead to frustration and feelings of failure, shame, and anxiety.

Students who always think about what they "should" do often exhaust themselves by doing too much and worrying excessively. Exhaustion is another factor that leads to poor test-taking results.

Breaking the "should" habit means replacing "should" statements with positive comments about what you have accomplished and what you hope to reasonably accomplish in the future. For example, instead of saying, "I shouldn't have gone to the football game" or "I should have stayed home and studied," say, "I studied for two hours before the football game, and then I had a good time. Two hours was plenty of study time for a Science quiz. I need to have time for friends as well as studying. I concentrated while studying, and I think I did a good job. Even if I don't get a perfect score on the Science quiz, I know I will do pretty well, and I gave myself the opportunity to do my best."

2. **Control physical symptoms**

Changing your physical response to stress can help break the test anxiety cycle. Relaxing is difficult when facing a major challenge such as the CSAP, but there are many proven techniques that can help you calm down.

- **Relax the morning of the test**—Try to allow yourself time to relax the morning of the test. Engaging in some mild exercise, such as taking a walk, will relieve a lot of your physical stress. Some students may find that a workout the night before an exam makes them feel more relaxed and helps them sleep well. This is probably because the exercise distracts the student from the upcoming test. Also, intense exercise releases chemicals in the brain that cause you to feel calmer and happier. It may only take a quick walk around the block to help you relax and get your mind off your problems.

- **Listen to music**—Listening to music in the morning before a test also may be helpful for students. It probably doesn't matter what kind of music you listen to as long as it makes you feel good about yourself, confident, and relaxed.

- **Relaxation exercises**—Relaxation exercises are helpful to many students. Stress causes many physical changes in the body, including tenseness in all muscle groups, increased heart rate, and other physical symptoms. Learning simple exercises to feel less tens can also help break the test anxiety cycle.

Most exercises include tightening and releasing tension in your body as well as deep breathing. The purpose of all of these exercises is to distract you from the anxiety of an upcoming test and to allow your body to feel more loose and relaxed. These exercises can be completed while sitting at your desk, taking a test, or studying.

Try this simple relaxation exercise the next time you are tense. Sit upright in your chair, but allow yourself to be comfortable. Close your eyes and take four deep breaths in and out. When you get to the fourth breath, start breathing quietly but remain focused on your breathing. Start increasing the tension in your feet by squeezing your toes together tightly and then slowly releasing the pressure. Feel how relaxed your toes are feeling? Now tighten and release other muscle groups. Go from your legs to your stomach, to your shoulders, to your hands, and finally to your forehead. Squeeze and tighten your muscles and then relax them, all while focusing on your breathing.

Once you practice this strategy, you might be able to feel more relaxed in a matter of seconds. This would be a good strategy to use during tests when you feel yourself becoming unfocused and anxious.

Prepare For the Science CSAP and Change the Way You Behave

Preparation always reduces anxiety. Taking the Science CSAP seriously, trying to do well on practice tests, and making an effort in all your classes will help you feel more confident and relaxed about the CSAP. Learning test-taking strategies also can give you a feeling of power and control over the test. No feeling is worse than realizing you are not prepared. Going into a test without ever having reviewed the Science CSAP material, looked at test-taking strategies, or concentrated on your schoolwork is very much like jumping out of an airplane without a parachute. You would be foolish if you were not panicked. Looking at the Science CSAP as just one more reason to take school seriously will help your grades, attitude, and success on the test.

1. **Use mental preparation**
   Before the test, imagine in step-by-step detail how you will perform well and obtain a positive result. Several days before the test, think through the day of the test; repeat this as many times as you need. Imagine getting up in the morning, taking a nice shower, getting dressed in comfortable clothes, and listening to music on your way to school. Think about sitting in the testing room with a confident expression on your face. Imagine yourself remembering all of the strategies you read about in this book and learned in your classroom. Go through an imaginary test, step by step, practicing what you will do if you encounter a difficult question. You also should repeat the positive thoughts that should go through your head during the test. Preparation like this is key for reducing anxiety, as you already feel you have taken the test prior to ever having stepped in that testing room!

2. Don't feel alone

People feel more anxious when they feel alone and separate from others. Have you ever worried about a problem in your family or something going wrong at school? Things seem much worse when you are alone, but when you talk to someone who cares about you, you will find your problems soon seem less worrisome. Talk to your friends, parents, and teachers about your feelings. You will be surprised at the support you receive. Everyone has anxious feelings about tests. Having others understand your anxious feelings will help you accept yourself even more. Other people in your life also can give you suggestions about tests and can help you put the Science CSAP and other tests in perspective.

3. Congratulate yourself during the test

Students with test anxiety spend a lot of time putting themselves down. They have never learned to say good things about themselves or to congratulate themselves on successes. As you go through the Science CSAP, try to find ways to mentally pat yourself on the back. If you find yourself successfully completing an extended-response question, tell yourself you did a good job. When you finish reading a test item and feel you understand the information fairly well, remind yourself you are doing a good job in completing the Science CSAP. Paying attention to your successes, and not focusing on your failures, can greatly reduce test anxiety.

Test-Taking Strategies

Test-Taking Strategies: Powerful Tools for Test Success

Test taking is a skill. When faced with tests, many eighth-grade students believe those who do well on tests are either part of the genius crowd at school or they study all the time. Sure, every middle school has its group of super students, and studying is certainly better than not studying, but the truth of the matter is learning to be an effective test taker is an important part of student performance. It's fine to have knowledge, but you also need to have the ability to show what you know. There are people with minds like encyclopedias who have a difficult time answering test questions correctly. All students can succeed on tests by taking school seriously and improving their test-taking strategies.

When students use test-taking strategies, they are sharpening the way they think so they can show what they know on tests. Using test strategies is not cheating, and it is not an easy way out. No test strategy is going to replace paying attention in school, working hard in your classes, and learning basic reading, writing, science, social studies, and CSAP skills. However, feeling comfortable and confident in using test-taking strategies can make a huge difference in test-taking success for many students on the Science CSAP as well as other tests students face throughout their schooling.

Take Care of Yourself

There are many things you can do to get your body ready for test day. If your body is ready, there is one less thing you have to think about, and you can focus on showing what you know.

- ***Eat right and don't skip breakfast***—If there is anything parents nag students about, it is getting enough sleep and eating right. Staying away from candy and chips is not going to make you a genius, but eating an appropriate breakfast on the morning of the Science CSAP will help more than you realize. (In fact, eating a reasonable breakfast every day will help you in your school work overall.) Eating a good breakfast does not mean eating a large breakfast or making anything complicated or time consuming. Most important is having a breakfast high in protein and low in sugar. Sugar quickly raises blood sugar levels, which then drop quickly, causing a student to feel tired and irritable and to lose concentration. Protein and foods lower in sugar provide more constant energy over the course of the day. Instead of going without breakfast or eating a bowl of sugary cereal, think about eating toast with natural peanut butter. This tastes good and is good for you, too. Remember, you will not do your best if you are hungry or have low blood sugar.

- ***Get to bed on time***—Getting to bed on time is a problem for most students. School starts quite early in the morning, and most eighth graders like to stay up late. Try to get yourself into a different sleep schedule a few days before the Science CSAP. For example, if you usually stay in your room listening to music until 11:30 p.m., try turning off your CD or MP3 player at 11:15 p.m. seven days before the test. Turn it off at 11:00 p.m. on the sixth day and fifth day before the test, 10:45 p.m. on the fourth day and third day before the test, and try to get to sleep not much later than 10:00 p.m. on the two days before the Science CSAP. You will be surprised at the difference a couple of hours of extra sleep makes in your performance.

- ***Dress for comfort, not for style, on test day***—Comfort is a must! Everybody wants to look "cool in school," but the day of the Science CSAP is not the time to wear clothes that are uncomfortable. You want to look nice and feel good about yourself, but don't wear that itchy sweater, tight pair of pants, or new pair of shoes.

Formulate a Plan

Any challenge is best met by a well thought-out plan of action. While some plans are better than others, looking at the whole picture before you proceed and thinking about the steps you will take to solve a problem or to respond to a challenge will help lead to success.

Planning is an important test-taking strategy. You may not realize it, but you create simple plans on a daily basis. For example, you might think, "I am going to have a busy day tomorrow, so I will make sure to leave enough time at lunch to go to the school store to buy what I need for my science project." If you plan, you will have time to do what you need and show what you know in science class. Many people make plans to solve problems, even though they might not think they are actually creating plans.

Successful test taking also requires a plan. Successful students do not just open up the test booklet and "go for broke." Instead, they take a few minutes to survey the test and get a general idea as to how they will tackle this challenge.

- *Begin by looking over the entire test*—Open your test booklet and briefly look at every page. How many pages are there? How many questions? Knowing this basic information can help you with your plan.

- *Read all directions*—Before you start answering test items, read all the directions given in your CSAP booklet. This is a very important step. The directions tell you what you're supposed to do. If you rush in without paying attention to these details, you could make careless mistakes. By reading directions, your task becomes clear in your head. Once you know what you're expected to complete, you can use your test-taking strategies to succeed.

- *Read all test information carefully*—Reading the directions is important, but so is reading each scenario, each chart or diagram, each question, and all the answer choices. Particularly with multiple-choice items, you may be tempted to stop reading the choices when you think you've found the correct answer. Don't stop; read each choice. You might find your first choice was wrong. By surveying all the information, you make the best possible selection. If you go with your first choice without reading all the choices, you make a decision without knowing all the facts.

- *Ignore information and details that are not important*—Pay careful attention to details given in both the question and the answer choices. Some questions may contain irrelevant information, but reading all the information gives you the best chance of success.

- *Pay attention to all test information*—Reading through the details gives you all the material you need to make all the best choices. You will be on your way to test-taking success!

Do Not Get Stuck on One Question

The Science CSAP was developed to allow students enough time to read and to answer all of the questions without feeling rushed. Almost every student should have enough time to carefully consider each question. If you survey the Science CSAP and plan out how you will use the available time, you will be able to finish the test without feeling rushed or stressed.

The Science CSAP was not designed to measure how fast you can answer questions. However, it is still important not to get stuck on a particular question. What causes the feeling of getting stuck? Some possible reasons include:

- you feel as though you have never seen the material before,
- you read and reread, but none of the choices seem correct,
- your mind suddenly goes blank, and
- you temporarily forget a fact or procedure you thought you knew.

These stuck feelings can oftentimes lead to panic. This can interfere with your performance on the rest of the test. If you get stuck on one question, you might not have enough time to finish all the questions on the test.

When you feel yourself becoming frustrated with a particular question, circle the number of that question and move on to another item. Keep in mind that no student is expected to answer all questions on the CSAP correctly. Missing a few questions is normal. If you waste time trying to solve a frustrating problem, you will have less time to complete the rest of the test successfully.

Many students find that when they come back to frustrating questions later on in the test, they are suddenly able to find the correct answers. This does not happen by luck or magic. Moving on from a stuck question and then coming back to that question later can result in success for two reasons:

- ***Remove yourself from the stressful situation***—Removing oneself from a stressful situation (the stuck question) and then finding success on other items can reduce a student's anxiety and boost self-confidence. Studies have shown that feeling calm and sure of oneself can be a big boost for success on tests and in schoolwork.

- ***The mind works in amazing ways***—By moving on to other test questions, you may trigger your mind into remembering information. Have you ever forgotten something, like someone's name? Imagine it's the first day at your new after-school job. The assistant manager walks in and you cannot remember her name. This is embarrassing. After a few hours, your mom's best friend, Ms. Garza, comes into the restaurant. Suddenly, your mind awakens! The supervisor's name is Ms. Garcia! How did you remember this? Ms. Garza's name begins with "G" and ends with "A." Recognizing Ms. Garza triggered your memory to remember your supervisor's name. On tests, you might find certain questions or answers will trigger your memory to other questions you thought were impossible.

Pay Attention to Yourself

When you become stuck on a question, are worried about time, or are concerned about how you are doing, it is easy to become curious about how others around you are doing. Looking around the room, thinking about the progress of other students, and letting your mind wander is a waste of your time. It is much more productive to pay attention to yourself than to others.

Take a Little Break

While it's important not to let your mind wander too far off track, it's OK to take short breaks if necessary. If you feel you are getting tense or worried, or find yourself becoming tired, give yourself a break between test questions. Perhaps you've finished a long constructed-response answer. It's OK to sit for a minute before moving on. You may want to stretch your arms and legs, think a positive thought, put down your pencil and pick it up again, close your eyes and concentrate, or do something that will reduce your anxiety and make you feel more calm or restful. You may find by moving around in your chair, you will "get your blood flowing again" and increase your concentration and focus on the questions to come.

Always, and This Means Always, Recheck Your Work

Careless errors can get in the way of doing well on a test. Even for students who feel fairly confident about tests, a test can present a stressful situation. The more stress you feel, the greater the chance for careless errors. Students sometimes think rechecking work takes too much time in a test situation and they could better use that time answering other questions. Actually, when you learn simple ways to recheck your work, it hardly takes any time at all. Studies have shown that rechecking work takes only about 5% of the total test time. Even if you only find one or two errors, this is a worthwhile investment of your efforts.

Here are easy ways to recheck your work:

- ***Look at the neatness of your work***—Neatness is especially important for constructed-response questions. For constructed-response questions, your written work will count as part of the answer, so make sure it is readable. But even on multiple-choice questions, you should make sure you have filled in all circles completely.

- ***Ask yourself, "Does my answer make sense?"***—On the Science CSAP, reread the question. Does your response answer the question being asked? If the question asks you for information about nuclear energy, make sure you didn't use an example about hydroelectric energy in your response. By reviewing items a second time, you may find a few errors.

Control Your Anxiety

Every student feels some degree of anxiety while taking a test. For some students, this anxiety helps them focus and gets them "psyched up" for a test-taking challenge. For other students, anxiety gets in the way of doing their best.

The Test Anxiety chapter in this book provides ways of dealing with stress during tests. If you feel that anxiety is a stumbling block on your way to test success, it is important that you reread the Test Anxiety chapter carefully and practice stress-reducing techniques.

What Kind of Questions are on the Science CSAP Assessment?

On the Science CSAP Assessment there are two types of questions, or items, to test your abilities. They are:

Multiple-Choice Items: You will be presented with four answer choices.

Constructed-Response Items: Constructed-response questions may ask you for a short response or extended response. These types of questions have a value of 1–4 score points and you can receive full or partial credit. You should try to answer these questions even if you are not sure of the correct answer. If a part of the answer is correct, you may get a portion of the points.

- **Short Constructed-Response Questions:** This type of question requires a response of a few sentences. You may not need to respond with a full paragraph.

- **Extended Constructed-Response Questions:** This type of question will require you to respond with more detail, with as many as four pieces of information or examples. You may need to create a logical paragraph, table, graph, or picture to answer the question.

Multiple-Choice Questions

Use "Codes" to Make Better Guesses

You might find it helpful to use "codes" to rate multiple-choice answer choices. Using your pencil in the test booklet, you can mark the following codes beside each multiple-choice response to see which is the best choice.

An example of a code used by an eighth-grade student is given below.

- (+) Put a "plus sign" by an answer choice if you are not sure if it is correct, but you think it might be correct;

- (?) Put a "question mark" by an answer choice if you are not sure if it is the correct answer, but you don't want to rule it out completely;

- (−) Put a "minus sign" by an answer choice if you are sure it is the wrong answer. (Then choose from the other answers to make an educated guess.)

Remember, it is fine to write in your test booklet. The space in the booklet is yours to use to help you do better on the CSAP. You will not have points counted off for using this coding system or creating your own system to help you on multiple-choice questions.

Answer Every Question

It is very important to answer as many multiple-choice questions as possible, even if you make an educated guess. On multiple-choice questions, you have a one in four chance of getting a question right—even if you just close your eyes and guess! This means that for every four questions you guess, the odds are you will get about one (25%) of the answers right. Guessing alone is not going to make you a star on the CSAP, but leaving multiple-choice questions blank is not going to help you either.

Take Advantage of "Chance"

On the Science section of the CSAP, it is very important to answer as many multiple-choice items as possible, even if you make a well thought-out guess, because luck is with you! If you can eliminate even one possible answer, your chances of success are now even better! The best way to improve your chances on multiple-choice items is to use strategies such as using codes and power guessing that are described in this chapter. Learning how to improve your chances by using educated guessing is not cheating. In fact, you probably use this strategy outside of the classroom and don't even think about it. Imagine you have misplaced your favorite CD, and you want to find it before you leave for your friend's house. There are many possible places that it could be, but you use your common sense to eliminate some possibilities, thereby saving time searching and increasing your chances of finding it in time. For example, it might be possible that you left it in your sister's room, but you remember, "That isn't likely because her CD player has been broken for a month." That leaves you one less place to look, and more chances for success.

Understand Multiple-Choice Questions and Answer Choices

Each multiple-choice item will have four possible answers that follow either a question or a statement. For example, you might be asked, "What process can change igneous rock into sedimentary rock?" You will be given four answer choices from which to choose the correct answer. Similarly, you could also be asked, "Within the rock cycle, igneous rock can change into sedimentary rock. What causes this change?" Possible answer choices could be "weathering," "melting," "heat," or "pressure." Not all questions will ask you to explain your answer. Even if the answer is as simple as "weathering," it is assumed that you used scientific reasoning to figure out a correct answer.

Answer choices are not designed to be tricky, but they won't be easy to choose correctly by guessing. Answer choices will not have one answer that is obviously incorrect. For example, if you are asked to choose the densest liquid in a chart that lists "water-1.00 g/ml," "milk-1.03 g/ml," "gasoline-0.66 g/ml," and "corn syrup-1.38 g/ml," you won't find a choice such as "olive oil-0.918 g/ml."

Answer choices will contain answers that might seem correct if you made a common mistake in reasoning or calculation. For example, if you are asked the percentage of offspring that will have brown eyes from a Punnet square, that calls for a different form of reasoning than if you were asked if the genotype of each parent is homozygous or heterozygous. There may be an answer choice that correctly reflects the genotype of each parent, but this would not be the correct answer to the original question regarding the percentage of brown-eyed offspring; it is distractor. Some answer choices also reflect common misunderstandings, such as finding the control variable when the question asks for the independent variable. When choosing answers, make sure to:

- *carefully read the question,*
- *check for any common or careless errors you may have made, and*
- *recheck your thinking and your calculations.*

Talk to Yourself!

Believe it or not, talking to yourself is a great Science CSAP strategy! You may not think about it, but you probably talk to yourself all the time when you are solving problems in everyday life, especially problems that have steps. Much of science calls for linear thinking, meaning that problems are best solved when certain steps are followed in order.

Imagine that you are having a problem with your computer. You are typing an essay that is due the next day, when suddenly your mouse will not work. If you are a successful problem solver you will first identify the problem: "The mouse isn't working." You then go through a checklist of how to proceed. "Let me see if there is a program running that is messing up my word processing." You hit a few keys and don't see any interference. The next step is to look at the wireless mouse. You say to yourself, "Let me see if there is a problem with the mouse." You pick it up and it looks fine. Then you say, "Well, something on my desk could be the problem." And you are right! Your big metal stapler is blocking the signal between your mouse and the computer. You solved the problem, step by step, by talking to yourself.

The same strategy works for science. In her biology lab, Keisha was asked to observe and describe the cell wall of an onion plant cell. Suppose she was asked, "What materials or instruments would she need to complete this assignment?" Identify the question, "This question has something to do with completing a science experiment." Then work through the processes involved, "I need to find what materials Keisha would need to observe the cell wall of an onion plant cell." Then ask yourself, "What instruments are typically used to observe and describe something in an experiment?" From that point on, you can eliminate and narrow down the instruments and materials that Keisha may need. For example, you can make a list of instruments you have used in Science class: a test tube, a beaker, a pipette, a hot plate, saline solution, iodine, a scale, a magnifying glass, a microscope, a glass slide, methylene blue, a scalpel, alcohol, and dissecting pins. Go through your list and cross out the instruments that you would not need to simply observe an onion plant cell. You should have a microscope, a glass slide, and methylene blue left on your list. Now, determine how each instrument or material may be used to complete Keisha's assignment: "A microscope will help Keisha visualize the tiny cells of the onion plant wall; a glass slide will help position the cell underneath the microscope for better viewing; methylene blue, a dye, will color the cell so that it can be viewed even more clearly underneath the microscope."

Learn How to "Power Guess"

Not everything you know was learned in a classroom. Part of what you know comes from just living your day-to-day life. When you take the Science CSAP, you should use everything you have learned in school, but you should also use your experiences outside of the classroom to help you correctly answer multiple-choice questions. You might think to yourself, "Well, science is different than other types of problems. There is just one correct answer. Things in science are either right or wrong. If I don't know the answer right away, nothing else will help me get the question right." Although you might think that science is different or harder, you still can use common sense thinking to help you do your best. Power guessing does not take the place of practice and knowledge, but it can help you to make reasonable choices by using what you know. Even if you eliminate one incorrect multiple-choice response, you have increased your chances of guessing correctly.

Constructed-Response Items

Answer every question

You are given several lines for constructed-response questions, which include short-answer and extended constructed-response items. A complete response not only utilizes the space given but also answers the question asked. Note, however, that incomplete but accurate and sensible answers may receive partial credit on the Science CSAP. It might be necessary to give a partial or an incomplete answer due to time limitations or not fully understanding the material. Giving a partial response is always better than not responding at all. Just as you will not lose points for incorrectly answering multiple-choice questions, you also will not lose points for making an attempt to answer a short-answer or extended-response item on the Science CSAP.

Write like a scientist

Expressing ideas about science can be different than other types of writing. Science writing is specific. For example, when talking to your friends and describing your car troubles, you might say, "The more I drove, stuff just got louder and louder." The scientific way is to identify the subject of your comments and conclusions and to use correct terms of measurement and other vocabulary to be as precise and as clear as possible. A more scientific way of explaining your problem might be, "As the mileage increased on the car, the noise volume of the engine increased as well." Using "it" should also be avoided. For example, when describing the data on a graph, it is appropriate to write, "The temperature of the soil becomes warmer as the number of decomposers increase," instead of "It gets warmer as the number of decomposers increase." Tell the reader what "it" is. Reviewing and using scientific vocabulary can also help you give the best possible answers on the Science CSAP.

Carefully look at the required response

In most instances, the type of response that is required will be made clear in the question. In order to not waste time, and to give the best possible answer, make sure to carefully understand how you are asked to respond. The answer space will lay out the way to respond. For example, there may be a blocked space for a diagram, and then a lined space for the definition or explanation. Ignoring what is in front of you is not a good idea.

- Suppose you review a diagram describing the manner in which metabolic energy was created. The short constructed-response question might be, "In the space below, write down the word that describes the unit of energy used to measure metabolism in humans. Then, write a sentence that explains a situation in which this measure is helpful." You should not leave the one word answer blank. You also know that you need to use one example to explain or define that measure. (The answer is "calorie," and it could be used by experts in sports physiology to determine what and when athletes eat before a competition.)

Don't forget to label

In extended constructed-response items, you might be asked to create a diagram, graph, or picture to explain your answer. Although you might know what you mean, the reader cannot read your mind. Make sure that you label or identify each axis of a graph, the columns of a table, or the objects in a picture. Not only will this help the reader of your test, but as you label your work, you will have a chance to review your thinking and make sure that your conclusions make sense.

Detailed answers rule

It is hard to show what you know if you don't include details. Suppose you are asked to compare two experiments to determine which type of battery is best in wet conditions. One experiment tested three different battery types when submerged in a tub of water, and the other experiment tested the same battery types when soaked with a damp cloth. Stating, "The environments were not the same, so the experiments cannot be adequately compared," may be generally correct, but further explaining that water pressure is also a variable provides the type of detail that lets the reader of your answer understand your reasoning and mastery of science.

Don't let your handwriting be frightening

In constructed-response items, you will be asked to write answers. Most students usually rely on computers to write assignments, and neat handwriting isn't always the first thing on their minds. Remember, even "science geniuses" will not do their best on the Science CSAP for Grade 8 unless the scorer can read their answers. No matter how rushed you may feel, take your time with your handwriting. It doesn't have to win awards, but it has to be legible.

Use your pencil to your advantage

Don't hesitate to make an outline of your answer on the margins of your test booklet. Look at constructed-response items as you would an essay question and take some time to write down the structure of your response. Then, add details. Thinking first and writing later is always a good idea.

Be the teacher for the day!

Constructed-response answers require YOU to be the teacher! Constructed responses often ask that you explain your answer in a way that is complete and understandable. Ask yourself, "If I were explaining my answer to a friend, would he or she understand?" Express your answers in a step-by-step manner, explaining your reasoning as you proceed. Don't assume that your "student" knows everything or can read your mind.

Don't avoid constructed-response items

Constructed-response items can be intimidating for some students, especially if they feel that they are weaker in science than in other subjects. While there are fewer constructed-response items on the CSAP, your answers are important in helping you do well on the science test.

Keep in mind that even a partially correct answer will increase your score, and there is no penalty for trying. If you find yourself feeling anxious about these more complex questions, review the section of this book about Test Anxiety, or talk to your science teacher or school counselor.

Constructed-response questions aren't always "harder" questions

Constructed-response questions basically ask how you arrived at your answer. The actual science may not be complicated or advanced. Remember, the purpose of learning science is to help you solve problems in school and in the rest of your life. Some science will come without much thinking, such as knowing that Earth is round. However, when you have to figure out an answer in college, you will need to put on your "thinking cap" to solve the problem. So, don't shy away from constructed-response items because you tell yourself, "This has to be WAY too hard." As you read the question, start to think about how you would describe your thinking and your answer, and don't forget to use scientific vocabulary in your response.

Science

Introduction

The Science CSAP reflects what you should know and should be able to do in the eighth grade. The CSAP will assess your knowledge with multiple-choice and constructed-response items. The questions are not meant to confuse or trick you but are written so you have the best opportunity to show what you know about science.

This Science chapter in the *Show What You Know® on the CSAP for Grade 8 Science, Student Self-Study Workbook* contains the following:

- Colorado's Standards, Item Distribution, and Scoring Guides for the Colorado Student Assessment Program in Science for Grade 8.

- A Science Practice Tutorial with explanations of each Assessment Objective, sample responses, student strategies, and an answer key with in-depth analyses. The Practice Tutorial will help you practice your test-taking skills.

- Two full-length Science Assessments (Assessment One—Session One, Session Two, and Session Three; and Assessment Two—Session One, Session Two, and Session Three) with sample responses, correlation charts, a standards checklist, and answer keys with in-depth analyses.

Both the Science Practice Tutorial and the Science Assessments have been created to model the Colorado Student Assessment Program in Science for Grade 8.

Understanding Assessment Objectives

Standard

Indicates the broad knowledge and skills that all students should be acquiring in Colorado schools at grade level. Each standard is assessed every year.

Standard 3 Life Science: Students know and understand the characteristics and structure of living things, the processes of life, and how living things interact with each other and their environment. *(Focus: Biology—Anatomy, Physiology, Botany, Zoology, Ecology)* **Students know and can demonstrate understanding that:**

Benchmark

Tactical descriptions of the knowledge and skills students should acquire by each grade level assessed by CSAP (5, 8, 10) or by district assessments at all grade levels.

3.1 Classification schemes can be used to understand the structure of organisms.

Assessment Objectives

Specific knowledge and skills selected to be measured by CSAP for each grade level. Assessment Objectives are assessed on a cyclical basis.

3.1.a Identify physical characteristics used to classify vertebrates.

The Colorado Standard's **numbering system** identifies the Standard, the Benchmark, and the Assessment Objective. For example, in the number 3.1.a, the first number (3) stands for the Standard, the second number (1) for the Benchmark (when applicable), and the letter (a) for the Assessment Objective.

Note: The Colorado Standards are not intended to take the place of a curriculum guide, but rather to serve as the basis for curriculum development to ensure that the curriculum is rich in content and is delivered through effective instructional activities. The Standards are in no way intended to limit learning, but rather to ensure that all students across the state receive a good educational foundation that will prepare them for a productive life.

Colorado's Science Standards

Standard 1 Students apply the processes of scientific investigation and design, conduct, communicate about, and evaluate such investigations. **Students know and are able to:**

Benchmark 1.1 Ask questions and state hypotheses that lead to different types of scientific investigations *(for example: experimentation, collecting specimens, constructing models, researching scientific literature).*

Assessment Objectives

1.1.a Plan and design a scientific investigation that includes:
- developing a testable question
- researching scientific literature
- stating a hypothesis
- identifying the independent and the dependent variables
- designing a written procedure for a controlled experiment
- using an appropriate observation/measurement technique for data collection
- keeping all other conditions constant

1.1.b Identify the independent and dependent variables in a previously conducted scientific investigation on a specific topic.

1.1.c Identify different methods used to investigate scientific questions *(e.g., controlled experiments, collecting specimens, constructing models, researching scientific literature, etc.).*

Benchmark 1.2 Use appropriate tools, technologies and metric measurements to gather and organize data and report results.

Assessment Objectives

1.2.a Record and report data from a scientific investigation using the appropriate tool and metric units.

1.2.b Describe how different types of technologies are used in scientific investigations *(e.g., telescopes, computers, calculators, seismographs, satellites, microscopes, etc.).*

1.2.c Construct and use different types of visual methods *(e.g., data tables, bar and line graphs, diagrams, etc.)* to summarize and present data.

Benchmark 1.3 Interpret and evaluate data in order to formulate a logical conclusion.

Assessment Objectives

1.3.a Interpret and evaluate data/observations *(e.g., data tables, bar and line graphs, diagrams, written descriptions, etc.)* to formulate a logical conclusion.

1.3.b Use evidence to state if a hypothesis is supported or not supported.

1.3.c Make predictions based on experimental data.

Benchmark 1.4 Demonstrate that scientific ideas are used to explain previous observations and to predict future events *(for example: plate tectonics and future earthquake activity)*.

Assessment Objective

1.4.a Evaluate collected data/observations and explain the patterns seen in past, current, and future scientific phenomena *(e.g., plate tectonics, future earthquake activity, etc.)*.

Benchmark 1.5 Identify and evaluate alternative explanations and procedures.

Assessment Objectives

1.5.a Describe other reasonable explanations, using the same independent and dependent variable, for the resulting data or observations from an investigation.

1.5.b Recognize and/or explain that alternative experimental designs can be used to investigate the same testable question.

Benchmark 1.6 Communicate results of their investigations in appropriate ways *(for example: written reports, graphic displays, oral presentations)*.

Assessment Objective

1.6.a Recognize that there are several different ways to communicate the results of investigations *(e.g., it is good to keep written reports so that information is preserved over time; oral presentations given to a large group are best when accompanied by a visual presentation; data is best suited for certain types of visual displays—bar graphs, line graphs, tables, etc.)*, and they are each used at different times.

Standard 2 Physical Science: Students know and understand common properties, forms, and changes in matter and energy. (*Focus: Physics and Chemistry*) **Students know and can demonstrate understanding that:**

Benchmark 2.1 Physical properties of solids, liquids, gases and the plasma state and their changes can be explained using the particulate nature of matter model.

Assessment Objectives

2.1.a Describe the particulate model for solid, liquid, gas, and plasma including the arrangement, motion, and energy of the particles (*for example, a lit fluorescent light bulb contains plasma which has widely spaced and highly energetic particles*).

2.1.b Using the kinetic molecular theory, predict how changes in temperature affect the behavior of particles of matter.

Benchmark 2.2 Mixtures of substances can be separated based on their properties (*for example: solubility, boiling points, magnetic properties, densities and specific heat*).

Assessment Objectives

2.2.a Explain how to use differences in solubility, boiling points, and magnetic properties to separate mixtures of substances (*for example, filtration can be used to separate mixtures by solubility or physical size*).

2.2.b Apply the concept of density to explain how mixtures of liquids and solids can be separated (*for example: relative densities—sinking and floating*).

Benchmark 2.3 Mass is conserved in a chemical or physical change.

Assessment Objectives

2.3.a Distinguish between a physical change and a chemical change.

2.3.b Apply the law of conservation of mass to physical changes (*for example, predict the mass of a substance after a phase change*).

2.3.c Apply the law of conservation of mass to chemical changes (*for example, determine the mass of products given the mass of reactants*).

Benchmark 2.4 Mass and weight can be distinguished.

Assessment Objectives

2.4.a Explain that the mass of an object is the amount of matter (*measured in grams using a balance*) it has and the weight of an object is the force of gravity (*measured in Newtons using a spring scale*) acting on its mass.

2.4.b Predict how changes in the force of gravity affect the mass and weight of an object (*for example: the mass of an object on the Moon will stay the same but its weight will be less than if the object were on Earth*).

Benchmark 2.5 All matter is made up of atoms that are comprised of protons, neutrons and electrons and when a substance is made up of only one type of atom, it is an element.

Assessment Objectives

2.5.a Identify that all matter is made up of atoms and that atoms are made of protons, neutrons, and electrons, and describe the location and charge of the parts of an atom.

2.5.b Identify that a substance made up of only one type of atom is an element, an atom is the smallest unit of an element that still retains the properties of that element, and different elements have different properties.

2.5.c Explain that the number of protons in an atom determines what element it is.

Benchmark 2.6 When two or more elements are combined a compound is formed which is made up of molecules.

Assessment Objectives

2.6.a Explain that two or more atoms may chemically combine to form a molecule, and recognize that a molecule can be represented by a chemical formula that shows the ratio of atoms of each element in the molecule *(for example, H_2 and H_2O are molecules)*.

2.6.b Describe that two or more elements may chemically combine to form a compound that may have different properties than the elements.

2.6.c Explain how mixtures are different than compounds.

2.6.d Identify that the smallest unit of a compound that still retains the properties of that compound is a molecule.

Benchmark 2.7 Quantities *(for example, time, distance, mass, force)* that characterize moving objects and their interactions within a system *(for example, force, speed, velocity, potential energy, kinetic energy)* can be described, measured and calculated.

Assessment Objectives

2.7.a Use measurements for objects that are moving in a straight line to relate distance, time, and average speed with words, graphs, and calculations.

2.7.b Identify the forces acting on a moving object and explain the effects of changes in the direction and magnitude of forces on the motion of the object.

2.7.c Compare the relative amount of potential energy *(stored energy)* and kinetic energy (energy of motion) of a moving object at different points along its path *(for example, a moving roller coaster has the most potential energy at the top of a hill and the most kinetic energy at the bottom of the hill)*.

Benchmark 2.8 There are different forms of energy and those forms of energy can be transferred and stored *(for example: kinetic, potential)* but total energy is conserved.

Assessment Objectives

2.8.a Recognize that energy is the ability to make objects move, and identify that mechanical, sound, thermal, solar, electromagnetic, chemical, and nuclear are some of the forms of energy.

2.8.b Explain that energy can be transferred *(moved)* from one object to another and transformed *(changed)* from one form to another.

2.8.c Identify the energy transformations that occur in a specific system.

2.8.d Apply the law of conservation of energy to describe what happens when energy is transferred and/or transformed.

Benchmark 2.9 Electric circuits provide a means of transferring electrical energy when heat, light, sound, magnetic effects and chemical changes are produced.

Assessment Objectives

2.9.a Describe the flow of electrons through a circuit.

2.9.b Identify series circuits and parallel circuits, and compare the two types of circuits.

Benchmark 2.10 White light is made up of different colors that correspond to different wavelengths.

Assessment Objectives

2.10.a Describe that white light is made of different colors of light *(ROYGBIV)*.

2.10.b Compare the relative wavelengths of different colors of light *(for example: red light has a longer wavelength than blue light)*.

Standard 3 Life Science: Students know and understand the characteristics and structure of living things, the processes of life, and how living things interact with each other and their environment. (*Focus: Biology—Anatomy, Physiology, Botany, Zoology, Ecology*) **Students know and can demonstrate understanding that:**

Benchmark 3.1 Classification schemes can be used to understand the structure of organisms.

Assessment Objectives

3.1.a Identify physical characteristics used to classify vertebrates.

3.1.b Classify organisms by their physical characteristics (*e.g., using a key, accessing prior knowledge*).

Benchmark 3.2 Human body systems have specific functions and interactions (*for example: circulatory and respiratory, muscular and skeletal*).

Assessment Objectives

3.2.a Identify organs, organ systems and describe their functions.

3.2.b Explain the interaction of body systems.

Benchmark 3.3 There is a differentiation among levels of organization (*cells, tissues, and organs*) and their roles within the whole organism.

Assessment Objective

3.3.a Identify levels of organization within an organism.

Benchmark 3.4 Multicellular organisms have a variety of ways to get food and other matter to their cells (*for example: digestion, transport of nutrients by circulatory system*).

Assessment Objectives

3.4.a Describe the various processes that food undergoes to be absorbed by an organism's cells.

3.4.b Identify and compare ways various organisms transport nutrients and wastes (*open and closed circulatory systems, plant vascular systems, etc.*).

3.4.c Identify and compare ways various organisms exchange carbon dioxide and oxygen (*stomata, lungs, skin, gills, etc.*) with the environment.

Benchmark 3.5 Photosynthesis and cellular respiration are basic processes of life (*for example, set up a terrarium or aquarium and make changes such as blocking out light*).

Assessment Objectives

3.5.a Describe the processes of photosynthesis and cellular respiration.

3.5.b Describe the relationship between photosynthesis and cellular respiration within plants and between plants and animals (*for example, animals can only do cellular respiration, plants do both*).

Benchmark 3.6 Different types of cells have basic structures, components and functions (*for example: cell membrane, nucleus, cytoplasm, chloroplast, single-celled organisms in pond water, Elodea, onion cell, human cheek cell*).

Assessment Objectives

3.6.a Identify cellular organelles and their functions.

3.6.b Differentiate between animal and plant cells and single celled organisms.

Benchmark 3.7 There are noncommunicable conditions and communicable diseases (*for example: heart disease and chicken pox*).

Assessment Objective

3.7.a Classify conditions as communicable or noncommunicable and recognize the cause of communicable diseases.

Benchmark 3.8 There is a flow of energy and matter in an ecosystem (*for example: as modeled in a food chain, web, pyramid, decomposition*).

Assessment Objectives

3.8.a Examine and analyze the flow of energy and matter in a dynamic ecosystem (*e.g., sun to producer to consumer, roles and importance of different organisms*).

3.8.b Infer the number of organisms or amount of energy available at each level of an energy pyramid.

Benchmark 3.9 Asexual and sexual cell reproduction/division can be differentiated.

Assessment Objectives

3.9.a Differentiate between mitosis and meiosis.

3.9.b Relate the number of chromosomes to the final product of mitosis or meiosis.

Benchmark 3.10 Chromosomes and genes play a role in heredity (*for example, genes control traits, while chromosomes are made up of many genes*).

Assessment Objectives

3.10.a Describe the relationship between chromosomes, genes and traits and their role in heredity.

3.10.b Infer the traits of the offspring based on the genes of the parents (*including dominant, recessive traits and use of Punnet square diagrams*).

Benchmark 3.11 Changes in environmental conditions can affect the survival of individual organisms, populations, and entire species.

Assessment Objectives

3.11.a Describe several environmental factors that could limit the size of an organism's population.

3.11.b Describe the impact of humans on the environment and how that affects the survival of populations and entire species.

3.11.c Describe how organisms change in response to environmental factors.

Benchmark 3.12 Changes or constancy in groups of organisms over geologic time can be revealed through evidence.

Assessment Objective

3.12.a Compare and contrast evidence of past life from different epochs to existing organisms.

Benchmark 3.13 Individual organisms with certain traits are more likely than others to survive and have offspring.

Assessment Objective

3.13.a Evaluate the potential of an organism with specific traits to survive and reproduce in an environment.

Standard 4 Earth and Space Science: Students know and understand the processes and interactions of Earth's systems and the structure and dynamics of Earth and other objects in space. (*Focus: Geology, Meteorology, Astronomy, Oceanography*) **Students know and can demonstrate understanding that:**

Benchmark 4.1 Inter-relationships exist between minerals, rocks and soils.

Assessment Objectives

4.1.a Understand the three types of rocks *(igneous, sedimentary, metamorphic)* and the processes that formed them through the rock cycle.

4.1.b Understand the composition and relationships of rocks, minerals, and soil formation.

Benchmark 4.2 Humans use renewable and nonrenewable resources *(for example: forests and fossil fuels).*

Assessment Objectives

4.2.a Understand the differences between renewable and nonrenewable energy resources.

4.2.b Predict the advantages and disadvantages of using both types of energy resources *(renewable and nonrenewable)* and their sustainability.

Benchmark 4.3 Natural processes shape Earth's surface *(for example: landslides, weathering, erosion, mountain building, volcanic activity).*

Assessment Objective

4.3.a Explain why Earth's surface is always building up in some places and wearing and down in others *(types of erosion, types of deposition).*

Benchmark 4.4 Major geological events such as earthquakes, volcanic eruptions, and mountain building are associated with plate boundaries and attributed to plate motion.

Assessment Objective

4.4.a Understand plate boundaries, their movements, and the resulting geologic events.

Benchmark 4.5 Fossils are formed and used as evidence to indicate that life has changed through time.

Assessment Objective

4.5.a Describe methods of fossil formation.

Benchmark 4.6 Successive layers of sedimentary rock and the fossils contained within them can be used to confirm age, geologic time, history, and changing life forms of the Earth; this evidence is affected by the folding, breaking and uplifting of layers.

Assessment Objectives

4.6.a Interpret rock layers, including position *(concept of superpositioning)*, composition and fossil content to determine past conditions.

4.6.b Predict the change in rock layer sequence due to folding, breaking and uplifting.

Benchmark 4.7 The atmosphere has basic composition, properties, and structure *(for example: the range and distribution of temperature and pressure in the troposphere and stratosphere).*

Assessment Objective

4.7.a Identify all of the layers of the atmosphere, their order and the properties and individual characteristics that define them.

Benchmark 4.8 Atmospheric circulation is driven by solar heating *(for example: the transfer of energy by radiation, convection, conduction).*

Assessment Objective

4.8.a Explain that the Sun heats Earth via radiation that in turn heats the atmosphere via conduction and convection.

Benchmark 4.9 There are quantitative changes in weather conditions over time and space *(for example: humidity, temperature, air pressure, cloud cover, wind, precipitation).*

Assessment Objective

4.9.a Interpret weather data and the changes that occur over time *(graph, charts, weather maps).*

Benchmark 4.10 There are large-scale and local weather systems *(for example: fronts, air masses, storms).*

Assessment Objectives

4.10.a Use several pieces of evidence *(cloud observations, weather maps)* to identify causes of changes in weather and weather patterns *(weather moves west to east).*

4.10.b Identify the inter-relationship between large scale weather systems and local weather.

4.10.c Explain how Earth's surface features *(such as mountains, oceans)* affect local weather.

Benchmark 4.11 The world's water is distributed and circulated through oceans, glaciers, rivers, groundwater, and atmosphere.

Assessment Objective

4.11.a Explain the processes and relationships that connect elements *(all water sources)* of the water cycle.

Benchmark 4.12 The ocean has a certain composition and physical characteristics *(for example: currents, waves, features of the ocean floor, salinity, and tides).*

Assessment Objective

4.12.a Understand the composition and physical characteristics of oceans *(for example: temperature, salinity, wavelength, ocean floor, etc.).*

Benchmark 4.13 There are characteristics *(components, composition, size)* and scientific theories of origin of the Solar System.

> **Assessment Objectives**
>
> **4.13.a** Describe the parts *(planets, Sun, moons, asteroids, comets)* of the Solar System and their motions.
>
> **4.13.b** Compare and contrast the characteristics of the Sun, Moon and Earth.
>
> **4.13.c** Examine and explain the scientific theories on the formation of our Solar System, Earth, and Moon.

Benchmark 4.14 Relative motion, axes tilt and positions of the Sun, Earth, and Moon have observable effects *(for example: seasons, eclipses, moon phases)*.

> **Assessment Objectives**
>
> **4.14.a** Understand how the location of the Moon affects the phases of the Moon, eclipses, and the tides.
>
> **4.14.b** Understand how the tilt and motions of Earth results in days, years and seasons.

Benchmark 4.15 The universe consists of many billions of galaxies *(each containing many billions of stars)* and that vast distances separate these galaxies and stars from one another and from Earth.

> **Assessment Objective**
>
> **4.15.a** Describe the components of the universe in terms of galaxies, stars, and solar systems.

Benchmark 4.16 Technology is needed to explore space *(for example: telescopes, spectroscopes, spacecraft, life support systems)*.

> **Assessment Objective**
>
> **4.16.a** Understand the technologies needed to explore space and evaluate their effectiveness and challenges.

Science—Colorado's Standards CSAP Science for Grade 8

Standard 5 Students understand that the nature of science involves a particular way of building knowledge and making meaning of the natural world. **Students know and can demonstrate understanding that:**

Benchmark 5.1 A controlled experiment must have comparable results when repeated.

Assessment Objectives

5.1.a Identify a controlled factor in a scientific investigation.

5.1.b Explain that by repeating a controlled experiment, it should lead to comparable results.

5.1.c Identify and/or explain that evidence collected through repeated experiments cannot be accurately compared to previous experimental results, if conditions were not kept the same.

Benchmark 5.2 Scientific knowledge changes as new knowledge is acquired and previous ideas are modified *(for example: through space exploration)*.

Assessment Objective

5.2.a Identify and/or describe the reasons why scientific knowledge changes over time.

Benchmark 5.3 Contributions to the advancement of science have been made by people in different cultures and at different times in history.

Assessment Objective

5.3.a Recognize the concept of multicultural contributions to the advancement of science over time.

Benchmark 5.4 Models can be used to predict change *(for example: computer simulation, video sequence, stream table)*.

Assessment Objectives

5.4.a Recognize and/or describe that models can be used to obtain information about scientific processes and/or objects that may be difficult to study.

5.4.b Describe a model that would be appropriate to understand a scientific process and content.

5.4.c Explain that models are used to understand processes and predict change in many situations:
- where it may take several years to collect the data firsthand (e.g., *sea floor spreading, etc.*)
- where the event has already occurred and evidence has been lost or is limited (e.g., *asteroid impact, fossil record, etc.*)
- when a process is dangerous to study (e.g., *volcanoes, earthquakes, tornados, etc.*)
- when a process is very slow (e.g., *erosion, continental drift, rock cycle, climate change, etc.*)
- when the scale of size is difficult to replicate and makes observations difficult (e.g., *atoms, cells, solar system, etc.*)
- to make an abstract more understandable (e.g., *Newton's Laws and amusement park physics, etc.*)

Benchmark 5.5 There are interrelationships among science, technology and human activity that affect the world.

Assessment Objective

5.5.a Explain that human activity, including current scientific studies and technological advancements, can have both positive and negative effects on the natural world.

About the CSAP Science

The Grade 8 CSAP Science Assessment will test you on your understanding of five science content standards for Grade 8. This will include Scientific Investigations, Connection Between Scientific Disciplines, Physical Science, Life Science, Earth and Space Science, and Science and Technology relating to Human Activity. The Science Assessment is given in three sessions: Session One, Session Two, and Session Three. Each session is 65 minutes.

For the Science Assessment, there are two different types of questions: multiple choice and constructed response.

Item Distribution on the CSAP Science for Grade 8

There will be a total of 80–83 test items on the CSAP Science; On an 83 question test, 60 test items will be multiple-choice questions, and 23 constructed-response questions.

Scoring

On the CSAP for Grade 8 Science Assessment, each multiple-choice item is worth one point. Constructed-response items may be worth up to four score points. Standard 6 is combined with Standard 1, and Standard 5 is combined with Standards 2–4.

Total test score points—98–100.

About the Science CSAP

CSAP Science for Grade 8 Blueprint*

Science Standards	CSAP Score
Standard 1 Scientific Investigations	30%
Standard 2 Physical Science	20%
Standard 3 Life Science	20%
Standard 4 Earth and Space Science	20%
Standard 5 Science and Technology Relating to Human Activity	6%
Standard 6 Connections Between Scientific Disciplines	4%

* As of the printing of this book an updated Science Blueprint had not been released. Standard 6 is combined with Standard 1 for testing.

Your answers to each constructed-response question on the CSAP will be scored by trained readers. Here is a sample Science Rubric that the readers will use to grade your constructed responses.

Constructed-response items may be scored on a four-point scale. Here is a sample of a 4-point rubric*:

Four-Point Rubric for Constructed-Response Items

4 points

This response provides extensive evidence of the kind of interpretation called for in the item or question. The response is well-organized, elaborate, and thorough. It demonstrates a complete understanding of the whole work as well as how parts blend to form the whole. It is relevant, comprehensive, and detailed, demonstrating a thorough understanding of the science passage. It thoroughly addresses the important elements of the question. It contains logical reasoning and communicates effectively and clearly.

3 points

This response provides evidence that an essential interpretation has been made. It is thoughtful and reasonably accurate. It indicates an understanding of the concept or item, communicates adequately, and generally reaches reasonable conclusions. It contains some combination of the following flaws: minor flaws in reasoning or interpretation, failure to address some aspect of the item, or the omission of some detail.

2 points

This response is mostly accurate and relevant. It contains some combination of the following flaws: incomplete evidence of interpretation, unsubstantial statements made about the text, an incomplete understanding of the concept or item, lack of comprehensiveness, faulty reasoning, and/or unclear communication.

1 point

This response provides little evidence of interpretation. It is unorganized and incomplete. It exhibits decoding rather than reading. It demonstrates a partial understanding of the item but is sketchy and unclear. It indicates some effort beyond restating the item. It contains some combination of the following flaws: little understanding of the concept or item, failure to address most aspects of the item, or inability to make coherent meaning from text.

0 points

A Zero is assigned if the response shows no understanding of the item or if the student fails to respond to the item.

* This rubric is a sample Science rubric; as of the printing of this book a Science rubric has not been released.

Glossary

Absorb: To suck up or drink in (a liquid); soak up; to take up or receive by chemical or molecular action.

Adaptation: A change by which an organism becomes better suited to its environment.

Affect: To act on; produce an effect or change in.

Air: The mixture of gases, mainly nitrogen and oxygen, that forms Earth's atmosphere.

Air mass: A body of air extending hundreds or thousands of miles horizontally and sometimes as high as the stratosphere and maintaining nearly uniform conditions of temperature and humidity at any given level as it travels.

Air pressure: The pressure exerted by the atmosphere.

Amino acid: Of a class of about twenty organic compounds which form the basic constituents of proteins and contain both acid and amino groups.

Amount: A quantity or degree of something, considered as a unit or total.

Amount of time: The measurement of how long it takes to do something.

Amplitude: The maximum extent of a vibration or oscillation from the point of equilibrium.

Anatomy: The science of the shape and structure of organisms and their parts.

Apply: The skill of selecting and using information in other situations or problems.

Asexual reproduction: Reproduction without the fusion of gametes.

Astronomy: The science of celestial objects, space, and the physical universe.

Atmosphere: The envelope of gases surrounding Earth or another planet.

Atom: The smallest particle of a chemical element, consisting of a positively charged nucleus surrounded by negatively charged electrons.

Axis: An imaginary straight line around which an object, such as Earth, rotates.

Balance scale: An accurate device used to measure the weight of chemicals and other substances.

Binary fission: A method of asexual reproduction, involves the splitting of a parent cell into two approximately equal parts.

Biodiversity: The variability among living organisms on Earth, including the variability within and between species and within and between ecosystems.

Biology: The scientific study of living organisms.

Biosphere: The part of Earth and its atmosphere in which living organisms exist or that is capable of supporting life.

Boiling point: The temperature at which a liquid boils at a fixed pressure, especially under standard atmospheric conditions.

Bone: Hard tissue that forms the skeleton of the body in vertebrate animals.

Botany: The scientific study of plants.

Brain: The controlling center of the nervous system in vertebrates, connected to the spinal cord and enclosed in the cranium; a nervous-system center in some invertebrates that is functionally similar to the brain in vertebrates.

Carbohydrate: Any of a group of organic compounds that includes sugars, starches, celluloses, and gums and serves as a major energy source in the diet of animals. These compounds are produced by photosynthetic plants and contain only carbon, hydrogen, and oxygen, usually in the ratio 1:2:1.

Carcinogen: A cancer-causing substance or agent.

Cause: (v) To make something happen or exist or be the reason that somebody does something or something happens; (n) Something that, or somebody who, makes something happen or exist or is responsible for a certain result.

Glossary

Cell: The smallest unit of living matter capable of functioning independently.

Cell division: The process in reproduction and growth by which a cell divides to form daughter cells.

Cellular respiration: The series of metabolic processes by which living cells produce energy through the oxidation of organic substances.

Centimeter (cm): A metric unit of length equal to one hundredth of a meter.

Characteristic: A feature or quality that makes somebody or something recognizable.

Chart: A diagram or table displaying detailed information.

Chemistry: The branch of science concerned with the properties and interactions of the substances of which matter is composed.

Chloroplast: A structure in algal and green plant cells which contains chlorophyll and in which photosynthesis takes place.

Chromosome: A thread-like structure found in the nuclei of most living cells, carrying genetic information in the form of genes.

Circuit: A closed path followed or capable of being followed by an electric current.

Classification: The systematic grouping of organisms into categories on the basis of evolutionary or structural relationships between them; taxonomy.

Climate: The average weather or the regular variations in weather in a region over a period of years.

Cloud: A visible body of very fine water droplets or ice particles suspended in the atmosphere at altitudes ranging up to several miles above sea level.

Cohesion: The intermolecular attraction by which the elements of a body are held together.

Color: The property of objects that depends on the light that they reflect and that is perceived as red, blue, green, or other shades.

Communicable disease: A disease that can be transferred from one person to another.

Community: A group of interdependent plants or animals growing or living together or occupying a specified habitat.

Composition: The combining of distinct parts or elements to form a whole.

Compost: A mixture of decayed plants and other organic matter used by gardeners for enriching soil.

Compound: A pure, macroscopically homogeneous substance consisting of atoms or ions of two or more different elements in definite proportions that cannot be separated by physical means. A compound usually has properties unlike those of its constituent elements.

Conclude: (v) To form an opinion or make a logical judgment about something after considering everything known about it.

Conclusion: A decision made or an opinion formed after considering the relevant facts or evidence.

Condensation: The process by which atmospheric water vapor liquifies to form fog, clouds, or the like, or solidifies to form snow or hail.

Condensation, heat of: Heat liberated by a unit mass of gas at its boiling point as it condenses into a liquid.

Condense: (v) To lose heat and change from a vapor into a liquid, or to make a vapor change to a liquid.

Conduction: The transmission or conveying of something through a medium or passage, especially the transmission of electric charge or heat through a conducting medium without perceptible motion of the medium itself.

Conductivity: The ability or power to conduct or transmit heat, electricity, or sound.

Glossary

Conservation: A law that states that matter and/or energy in a closed system are constant.

Conserve: To use something sparingly so as not to exhaust supplies.

Consumer: In an ecological community or food chain, an organism that feeds on other organisms, or on material derived from them.

Continent: Any one of the seven large continuous land masses that constitute most of the dry land on the surface of Earth.

Controlled experiment: An experiment that isolates the effect of one variable on a system by holding constant all variables but the one under observation.

Convection: Heat transfer in a gas or liquid by the circulation of currents from one region to another.

Conversion: A change in the nature, form, or function of something.

Coriolis effect: Result of an apparent force that, as a result of Earth's rotation, deflects moving objects (as projectiles or air currents) to the right in the Northern Hemisphere and to the left in the Southern Hemisphere.

Cycle: A sequence of events that is repeated again and again.

Data: Information, often in the form of facts or figures obtained from experiments or surveys, used as a basis for making calculations or drawing conclusions.

Decomposer: An organism, especially a bacterium or fungus, that causes organic matter to rot or decay.

Decomposition: Breakdown or decay of organic materials.

Demonstrate: To show or prove something clearly and convincingly.

Density: The mass of a substance per unit volume.

Depend: To be affected or decided by other factors.

Describe: The skill of developing a detailed picture, image, or characterization using diagrams and/or words, written, or oral.

Design: The application of scientific concepts and principles and the inquiry process to the solution of human problems that regularly provide tools to further investigate the natural world.

Diagram: A simple drawing showing the basic shape, layout, or workings of something.

Diameter: A straight line running from one side of a circle or other rounded geometric figure through the center to the other side, or the length of this line; the width or thickness of something, especially something circular or cylindrical.

Direction: The management or control of somebody or something by providing instructions.

DNA (deoxyribonucleic acid): A substance which is present in the cell nuclei of nearly all living organisms and is the carrier of genetic information.

Earth: The third planet in order from the sun with an orbital period of 365.26 days, a diameter of 12,756 km (7,926 mi), and an average distance from the sun of 149,600,000 km (93,000,000 mi).

Earthquake: A violent shaking of Earth's crust that may cause destruction to buildings and installations and results from the sudden release of tectonic stress along a fault line or volcanic activity.

Echo: The repetition of a sound caused by the reflection of sound waves from a surface.

Eclipse: The partial or complete obscuring, relative to a designated observer, of one celestial body by another.

Ecology: The branch of biology concerned with the relations of organisms to one another and to their physical surroundings.

Ecosystem: A localized group of interdependent organisms together with the environment that they inhabit and depend on.

Glossary

Effect: The result or consequence of an action, influence, or causal agent.

Egg: A large sex cell produced by birds, fish, insects, reptiles, or amphibians, enclosed in a protective covering that allows the fertilized embryo to continue developing outside the mother's body until it hatches; a female reproductive cell.

Electrical: Caused by electricity or something that uses or conveys electricity.

Electricity: A fundamental form of kinetic or potential energy created by the free or controlled movement of charged particles, such as electrons, positrons, and ions.

Electromagnetic radiation: A kind of radiation, including visible light, radio waves, gamma rays, and x-rays, in which electric and magnetic fields vary simultaneously.

Electron: A stable, negatively charged subatomic particle with a mass less than that of the proton, found in all atoms and acting as the primary carrier of electricity in solids.

Element: A substance composed of atoms having an identical number of protons in each nucleus. Elements cannot be reduced to simpler substances by normal chemical means.

Elevation: Height above a given level, especially sea level.

Energy: The ability or power to work or make an effort.

Energy of motion (kinetic): The energy that a body or system has because of its motion.

Environment: The complex of physical, chemical, and biotic factors (as climate, soil, and living things) that act upon an organism or an ecological community and ultimately determine its form and survival.

Equilibrium: The state of a chemical reaction in which its forward and reverse reactions occur at equal rates so that the concentration of the reactants and products does not change with time.

Erode: To wear away outer layers of rock or soil, or to be gradually worn away by the action of wind or water.

Erosion: The gradual wearing away of rock or soil by physical breakdown, chemical solution, and transportation of material, as caused, for example, by water, wind, or ice.

Eruption: The violent ejection of material, such as gas, steam, ash, or lava from a volcano.

Evaporate: To change a liquid into a vapor, usually by heating to below its boiling point, or to change from a liquid to vapor in this way.

Evaporation: A process in which something is changed from a liquid to a vapor without its temperature reaching the boiling point.

Event: A happening or occurrence.

Evidence: Observations, measurements, or data collected through established and recognized scientific processes.

Experiment: A test under controlled conditions that is made to demonstrate a known truth, examine the validity of a hypothesis, or determine the efficacy of something previously untried.

Explain: The skill of making a theory, hypothesis, inference, or conclusion plain and comprehensible—includes supporting details with an example.

Explanation: The giving of details about something or reasons for something.

Feet: Plural for foot; one foot is a unit of length in the U.S. customary and British imperial systems equal to .3048 m (12 inches). There are three feet in a yard.

Flower: A colored, sometimes scented, part of a plant that contains the plant's reproductive organ.

Food: Material that provides living things with the nutrients they need for energy and growth.

Food chain: A hierarchy of different living things, each of which feeds on the one below.

Glossary

Food pyramid: A graphic representation of the structure of a food chain, depicted as a pyramid having a broad base formed by producers and tapering to a point formed by end consumers. Between successive levels, total biomass decreases as energy is lost from the system.

Food web: A complex of interrelated food chains in an ecological community.

Force: The power, strength, or energy that somebody or something possesses; a physical influence that tends to change the position of an object with mass, equal to the rate of change in momentum of the object: Symbol F.

Forest: A large area of land covered in trees and other plants growing close together, or the trees growing on it.

Fossil: The remains of an animal or plant preserved from an earlier era inside a rock or other geological deposit, often as an impression or in a petrified state.

Fossil fuel: A hydrocarbon deposit, such as petroleum, coal, or natural gas, derived from living matter of a previous geologic time and used for fuel.

Freeze: To be changed, or cause liquid to change, into a solid by the loss of heat, especially to change water into ice; to harden, or cause something to harden, through the effects of cold or frost.

Frequency: The number of complete cycles of a periodic process occurring per unit time.

Friction: The rubbing of two objects against each other when one or both are moving.

Front: The interface between air masses of different temperatures or densities.

Function: An action or use for which something is suited or designed.

Galaxy: A system of millions or billions of stars, together with gas and dust, held together by gravitational attraction.

Gas: A substance, such as air, that is neither a solid nor a liquid at ordinary temperatures and that has the ability to expand indefinitely.

Genetics: The branch of biology that deals with heredity, especially the mechanisms of hereditary transmission and the variation of inherited characteristics among similar or related organisms.

Geologic time: The period of time covering the physical formation and development of Earth, especially the period prior to human history.

Geology: The scientific study of the origin, history, and structure of Earth.

Geosphere: The solid part of Earth consisting of the crust and outer mantle.

Germinate: To start to grow from a seed or spore into a new individual; to be created and start to develop.

Glacier: A large body of continuously accumulating ice and compacted snow, formed in mountain valleys or at the poles, that deforms under its own weight and slowly moves.

Gram (g): A metric unit of mass, equal to 0.001 kg or equivalent to approximately 0.035 oz.

Graph: A diagram used to indicate relationships between two or more variable quantities. The quantities are measured along two axes, usually at right angles.

Grassland: Land on which grass or low green plants are the main vegetation.

Gravitation: A gradual and steady movement to or toward somebody or something as if drawn by some force or attraction; the mutual force of attraction between all particles or bodies that have mass.

Gravity: The attraction due to gravitation that Earth or another celestial body exerts on an object on or near its surface.

Greenhouse effect: The phenomenon whereby Earth's atmosphere traps solar radiation caused by the presence in the atmosphere of gases, such as carbon dioxide, water vapor, and methane, that allow incoming sunlight to pass through but absorb heat radiated back from Earth's surface.

Glossary

Greenhouse gas: A gas, such as carbon dioxide, that contributes to the greenhouse effect by absorbing infrared radiation.

Groundwater: Water beneath Earth's surface, often between saturated soil and rock, which supplies wells and springs.

Habitat: The area or environment where an organism or ecological community normally lives or occurs.

Hand lens: A magnifying glass with a handle for holding in the hand.

Hardness: The state or quality of being firm, solid, and compact; the degree to which a metal may be scratched, abraded, indented, or machined, measured according to any of several scales.

Heart: A hollow muscular organ that pumps blood around the body; in humans it is situated in the center of the chest with its apex directed to the left.

Heat energy: A form of transferred energy that arises from the random motion of molecules and is felt as temperature, especially as warmth or hotness.

Heat of condensation: Heat liberated by a unit mass of gas at its boiling point as it condenses into a liquid.

Homeostasis: The ability or tendency of an organism or cell to maintain internal equilibrium by adjusting its physiological processes.

Humidity: The amount of water suspended in the air in tiny droplets.

Hydrologic cycle: The cycle of evaporation and condensation that controls the distribution of Earth's water as it evaporates from bodies of water, condenses, precipitates, and returns to those bodies of water.

Hydrosphere: The watery layer of Earth's surface; includes water vapor.

Hypothesis: A tentative explanation for an observation, phenomenon, or scientific problem that can be tested by further investigation.

Identify: To recognize somebody or something and to be able to say who or what he, she, or it is.

Inch (in): A unit of length equal to 2.54 cm or 1/12th of a foot.

Infrared: Electromagnetic radiation having a wavelength just greater than that of red light but less than that of microwaves, emitted particularly by heated objects.

Inherited: To receive a characteristic or quality as a result of its being passed on genetically.

Input: The addition of matter, energy, or information to a system; a change of matter or energy in the system; a living organism learning something new.

Inquiry: The skill of the investigative process characterized by asking questions of the natural world, developing hypotheses, testing hypotheses by manipulating variables and measuring responding variables, and drawing inferences from data to develop correlations between variables or cause-effect relationships between variables.

Invent: To create something.

Invention: Something that someone has created, especially a device or process.

Invertebrate: An animal, such as an insect or mollusk, which lacks a backbone or spinal column.

Investigate: To carry out a detailed examination or inquiry in order to find out about something.

Investigation: A multifaceted and organized scientific study of the natural world that involves making observations; asking questions; gathering information through planned study in the field, laboratory, or research setting; and using tools to gather data that is analyzed to find patterns and is subsequently communicated.

Kinetic energy: Energy which a body possesses by virtue of being in motion.

Kilogram (kg): The basic unit of mass in the SI system, equal to 1,000 grams or 2.2046 lbs.

Kilometer (km): The basic unit of measure in the SI system, equal to 1,000 meters or 0.621 miles.

Glossary

Lake: A large body of water surrounded by land.

Leaf: Any of the flat green parts that grow in various shapes from the stems or branches of plants and trees; its main function is photosynthesis.

Learned (acquired) characteristic: A characteristic that an organism develops in response to its environment and cannot be passed on to the next generation.

Lever: A rigid bar that pivots about a point (fulcrum) and is used to move or lift a load at one end by applying force to the other end.

Life cycle: The course of developmental changes in an organism from fertilized zygote to maturity when another zygote can be produced.

Light: Electromagnetic radiation that can produce a visual sensation.

Liquid: A substance in a condition in which it flows, that is a fluid at room temperature and atmospheric pressure, and that has shape, but not volume, can be changed.

Liter (L): A unit of volume equal to 1 cubic decimeter or 1.056 liquid quarts.

Living: Alive, not dead.

Lithosphere: The rigid outer part of Earth, consisting of the crust and upper mantle.

Logical plan: An investigative plan that has coherence among all its attributes, including hypotheses, observations, and data to support the hypotheses, and logical inference to support conclusions.

Lung: In air-breathing vertebrate animals, either of the paired spongy respiratory organs, situated inside the rib cage, that transfer oxygen into the blood and remove carbon dioxide from it.

Machine: A device with moving parts, often powered by electricity, used to perform a task, especially one that would otherwise be done by hand.

Macromolecule: A very large molecule, such as a polymer or protein, consisting of many smaller structural units linked together.

Macroscopic: Large enough to be perceived or examined by the unaided eye.

Magnetic: Able to attract iron or steel objects.

Magnifying glass: A convex lens in a frame with a handle, used to make objects viewed through it appear larger.

Mass: The property of an object that is a measure of its inertia, the amount of matter it contains, and its influence in a gravitational field: Symbol m.

Material: Relating to or consisting of solid physical matter.

Matter: The material substance of the universe that has mass, occupies space, and is convertible to energy.

Meiosis: The process of cell division in sexually reproducing organisms that reduces the number of chromosomes in reproductive cells from diploid to haploid, leading to the production of gametes in animals and spores in plants.

Melt: To change a substance from a solid to a liquid state by heating it, or be changed in this way.

Meteorology: The science that deals with the phenomena of the atmosphere, especially weather and weather conditions.

Meter (m): The basic unit of length in the SI system, equivalent to approximately 1.094 yd or 39.37 in.

Microscopic: Too small to be seen by the unaided eye but large enough to be studied under a microscope.

Mile (mi): A unit of linear measurement on land equivalent to 5,280 ft or 1,760 yd or 1.6 km.

Milliliter (mL): A unit of volume equal to one thousandth of a liter.

Glossary

Mineral: An inorganic substance that must be ingested by animals or plants in order to remain healthy. For example, the minerals that a plant takes from the soil or the constituents in food that keep a human body healthy and help it grow.

Mineral: A naturally occurring, homogeneous inorganic solid substance having a definite chemical composition and characteristic crystalline structure, color, and hardness.

Mitosis: A type of cell division in which daughter cells have the same number and kind of chromosomes as the parent nucleus.

Mixture: A composition of two or more substances that are not chemically combined with each other and are capable of being separated.

Model: A representation of a system, subsystem, or parts of a system that can be used to predict or demonstrate the operation or qualities of the system.

Molecule: The smallest physical unit of a substance that can exist independently, consisting of one or more atoms held together by chemical forces.

Moon: Earth's only natural satellite; the astronomical body nearest to Earth, except for some artificial satellites and occasional meteors.

Moon (lunar) phases: One of the cyclically recurring apparent forms of the moon.

Motion: A natural event that involves a change in the position or location of something.

Mountain: A high and often rocky area of a land mass with steep or sloping sides.

Multicellular: Having or consisting of many cells.

Muscle: A tissue that is specialized to undergo repeated contraction and relaxation, thereby producing movement of body parts, maintaining tension, or pumping fluids within the body.

Mutation: A change in genetic structure which results in a variant form and may be transmitted to subsequent generations.

Natural resources: A material source of wealth, such as timber, fresh water, or a mineral deposit that occurs in a natural state and has economic value.

Natural selection: The process in nature by which only the organisms best adapted to their environment tend to survive and transmit their genetic characteristics in increasing numbers to succeeding generations while those less adapted tend to be eliminated.

Neutron: A subatomic particle of about the same mass as a proton but without an electric charge.

Newtons (N): The basic unit of force in the SI system, equivalent to the force that produces an acceleration of one meter per second on a mass of one kilogram.

Niche: The function or position of an organism or population within an ecological community.

Nonliving: Not containing or supporting life.

Nonrenewable resource: Of or relating to an energy source, such as oil or natural gas, or a natural resource, such as a metallic ore, that is not replaceable after it has been used.

Nutrient (mineral): Any substance that provides nourishment; for example, the minerals that a plant takes from the soil or the constituents in food that keep a human body healthy and help it grow.

Object: Something that can be seen or touched.

Observe: To watch someone or something attentively, especially for scientific purposes.

Observation: The skill of recognizing and noting some fact or occurrence in the natural world, including the act of measuring.

Ocean: A large expanse of salt water, especially any of Earth's five largest such areas: the Atlantic, Pacific, Indian, Arctic, and Antarctic oceans.

Oceanography: The branch of science concerned with the physical and biological properties and phenomena of the sea.

Glossary

Orbit: (n) The path that a celestial body, such as a planet, moon, or satellite, follows around a larger celestial body, such as the sun. (v) To move around a celestial body in a path dictated by the force of gravity exerted by that body.

Organ: A differentiated part of an organism, such as an eye, wing, or leaf, which performs a specific function.

Organism: A living thing, such as a plant, animal, or bacterium.

Organize: To arrange the elements of something in a way that creates a particular structure.

Ounce (oz): A unit of weight equal to one-sixteenth of a pound; a unit for measuring liquid equal to 0.0284 of a liter.

Output: The removal of matter, energy, or information from a system; a change of matter or energy in the system; a living organism produces and excretes a substance.

Oxygen: A colorless, odorless gas that is the most abundant chemical element and forms compounds with most others: Symbol O.

Pangaea (plate tectonics): A hypothetical supercontinent that included all the landmasses of Earth before the Triassic Period. When continental drift began, Pangaea broke up into Laurasia and Gondwanaland.

Parasite (parasitic): An organism that grows, feeds, and is sheltered on or in a different organism while contributing nothing to the survival of its host.

Part: Any of several equal portions that make up something, such as a mixture.

Pattern: A regular or repetitive form, order, or arrangement.

Periodic table: A table of the chemical elements arranged in order of atomic number, usually in rows, with elements having similar atomic structure appearing in vertical columns.

pH p(otential of) H(ydrogen): A measure of the acidity or alkalinity of a solution, numerically equal to 7 for neutral solutions, increasing with increasing alkalinity and decreasing with increasing acidity. The pH scale commonly in use ranges from 0 to 14.

Photosynthesis: The process in green plants and certain other organisms by which carbohydrates are synthesized from carbon dioxide and water using light as an energy source. Most forms of photosynthesis release oxygen as a byproduct.

Physical change: A change from one state (solid or liquid or gas) to another without a change in chemical composition.

Physics: The science of matter and energy and of interactions between the two.

Physiology: The branch of biology concerned with the normal functions of living organisms and their parts.

Pitch: The level of a sound in the scale, defined by its frequency.

Plan: A method of doing something that is worked out usually in some detail before it is begun and that may be written down in some form or simply retained in memory.

Planet: An astronomical body that orbits a star and does not shine with its own light, especially one of the eight such bodies orbiting the sun in the solar system.

Plasma: A phase of matter distinct from solids, liquids, and normal gases.

Plate tectonics: A theory that explains the global distribution of geological phenomena, such as seismicity, volcanism, continental drift, and mountain building, in terms of the formation, destruction, movement, and interaction of Earth's lithospheric plates.

Plateau: An elevated, comparatively level expanse of land.

Polarity: The state of having poles or opposites.

Glossary

Pollute: To cause harm to an area of the natural environment, for example, the air, soil, or water, usually by introducing damaging substances, such as chemicals or waste products.

Pollination: Transfer of pollen from the anther to the stigma of a plant.

Population: All the organisms that constitute a specific group or occur in a specified habitat.

Potential energy: Energy possessed by a body by virtue of its position or state.

Pound (lb): A unit of weight divided into 16 ounces and equivalent to 0.45 kg.

Precipitation: Rain, snow, or hail, all of which are formed by condensation of moisture in the atmosphere and fall to the ground.

Predict: To say what is going to happen in the future, often on the basis of present indications or past experience.

Prediction: The skill of predicting a future event or process based on theory, investigation, or experience.

Pressure: Force applied uniformly over a surface, measured as force per unit of area.

Prevailing wind: A wind from the predominant or most usual direction.

Problem: A question or puzzle that needs to be solved.

Procedure: An established or correct method of doing something.

Process: A series of actions directed toward a particular aim; a series of natural occurrences that produce change or development.

Producer: An organism, such as a green plant, that manufactures its own food from simple inorganic substances.

Property: A characteristic quality or distinctive feature of something (often used in the plural form: properties).

Properties: The basic or essential attributes shared by all members of a group.

Proton: A stable subatomic particle occurring in all atomic nuclei, with a positive electric charge equal in magnitude to that of an electron.

Pull: To apply force to a physical object so as to draw or tend to draw it toward the force's origin.

Pulley: A mounted rotating wheel with a grooved rim over which a belt or chain can move to change the direction of a pulling force.

Push: The act of applying pressure or force to somebody or something in order to move that person or object.

Question: A problem to be discussed or solved in an examination or experiment.

Radiation: Energy emitted as electromagnetic waves or subatomic particles.

Radiometric dating: A method of determining the age of objects or material using the decay rates of radioactive components, such as potassium-argon.

Radius: A straight line extending from the center of a circle to its edge or from the center of a sphere to its surface: Symbol r.

Rate: The speed at which one measured quantity happens, runs, moves, or changes compared to another measured amount, such as time.

Recycle: To process used or waste material so that it can be used again.

Reduce: To become or make something smaller in size, number, extent, degree, or intensity.

Renewable resource: Any natural resource (such as wood or solar energy) that can be replenished naturally with the passage of time.

Replication: The process whereby DNA makes a copy of itself before cell division.

Report: To give detailed information about research or an investigation.

Glossary

Reproduce: To produce offspring or new individuals through a sexual or asexual process.

Reproduction: The production of young plants and animals of the same kind through a sexual or asexual process.

Result: To produce a particular outcome.

River: A natural formation in which fresh water forms a wide stream that runs across the land until it reaches the sea or another area of water.

RNA (ribonucleic acid): A substance in living cells which carries instructions from DNA for controlling the synthesis of proteins and in some viruses carries genetic information instead of DNA.

Rock: Any natural material with a distinctive composition of minerals.

Root: The part of a plant that has no leaves or buds and usually spreads underground, anchoring the plant and absorbing water and nutrients from the soil.

Rotation: The act or process of turning around a center or an axis.

Salinity: The relative proportion of salt in a solution.

Satellite: Any celestial body orbiting around a planet or star.

Scavenger: An animal, bird, or other organism that feeds on dead and rotting flesh or discarded food scraps.

Science: The systematized knowledge of the natural world derived from observation, study, and investigation; also the activity of specialists to add to the body of this knowledge.

Scientific: Relating to, using, or conforming to science or its principles; proceeding in a systematic and methodical way.

Scientific law: A phenomenon of nature that has been proven to invariably occur whenever certain conditions exist or are met.

Scientific theory: A well-substantiated explanation of some aspect of the natural world; an organized system of accepted knowledge that applies in a variety of circumstances to explain a specific set of phenomena; "scientific theories must be falsifiable."

Scientist: Someone who has had scientific training or who works in one of the sciences.

Sea: The great body of salt water that covers a large portion of Earth.

Season: One of the natural periods into which the year is divided by the equinoxes and solstices or atmospheric conditions.

Sediment: Material eroded from preexisting rocks that is transported by water, wind, or ice and deposited elsewhere.

Sedimentary: Rock that has formed from sediment deposited by water or wind.

Seed: The body, produced by reproduction in most plants, that contains the embryo and gives rise to a new individual. In flowering plants, it is enclosed within the fruit.

Sexual reproduction: Reproduction by the union or fusion of two differing gametes.

Shadow: Relative darkness in a place that is being screened or blocked off from direct sunlight.

Shape: The outline of something's form.

Size: The amount, scope, or degree of something, in terms of how large or small it is.

Skeleton: The rigid framework of interconnected bones and cartilage that protects and supports the internal organs and provides attachment for muscles in humans and other vertebrate animals.

Skepticism: The attitude in scientific thinking that emphasizes that no fact or principle can be known with complete certainty; the theory that all knowledge is uncertain.

Glossary

Soil: The top layer of most of Earth's land surface, consisting of the unconsolidated products of rock erosion and organic decay, along with bacteria and fungi.

Solar: Relating to or originating from the sun.

Solid: Consisting of compact unyielding material having no open interior spaces; not hollow.

Solubility: The quality or condition of being soluble.

Soluble: That can be dissolved, especially easily dissolved.

Solutions: Artifacts of the scientific design process in response to human problems that can include devices or processes, such as environmental impact statements.

Solve: To find a way of dealing successfully with a problem or difficulty.

Sort: To arrange data in a set order.

Sound: Vibrations traveling through air, water, or some other medium, especially those within the range of frequencies that can be perceived by the human ear.

Space: The expanse in which the solar system, stars, and galaxies exist; the universe.

Special: Designed or reserved for a particular purpose.

Species: A fundamental category of taxonomic classification, ranking below a genus or subgenus and consisting of related organisms capable of interbreeding.

Specific heat: The ratio of the amount of heat required to raise the temperature of a unit mass of a substance by one unit of temperature to the amount of heat required to raise the temperature of a similar mass of a reference material, usually water, by the same amount.

Spectroscope: An instrument for producing and observing spectra, the entire range of wavelengths of electromagnetic radiation.

Speed: The rate at which something moves, happens, or functions.

Spin (rotate): To turn or make something turn round and round rapidly, as if on an axis.

Spring scale: A balance that measures weight by the tension on a spiral spring.

Sprout: To begin to grow from a seed; a new growth on a plant; for example, a bud or shoot.

Star: A celestial body of hot gases that radiates energy derived from thermonuclear reactions in the interior.

States of matter: The three traditional states of matter are solids, liquids, and gases.

Stem: The main axis of a plant that bears buds and shoots.

Storm: A violent disturbance of the atmosphere with strong winds and usually rain, thunder, lightning, or snow.

Stratosphere: The atmospheric layer between the troposphere and the mesosphere.

Stream: A narrow and shallow river; a current of air or water.

Strength: The ability to withstand force, pressure, or stress.

Structure: A part of a body or organism; for example, an organ or tissue, identifiable by its shape and other properties.

Substance: A particular kind of matter or material.

Summary: A shortened version of something that has been said or written, containing only the main points.

Sun: The star at the center of our solar system around which Earth and the seven other planets orbit. It provides us with heat and light.

Symbiotic: A close, prolonged association between two or or more organisms.

Glossary

Synthesis: Formation of a compound from simpler compounds or elements.

System: An assemblage of interrelated parts or conditions through which matter, energy, and information flow.

Table: An arrangement of information or data into columns and rows or a condensed list.

Telescope: A scientific instrument designed to collect and record electromagnetic radiation from cosmic sources.

Temperature: The heat of something measured on a particular scale, such as the Fahrenheit or Celsius scale: Symbol T.

Texture: The feel and appearance of a surface, especially how rough or smooth it is.

Thaw: To melt or make something melt.

Theory (scientific): A well-substantiated explanation of some aspect of the natural world; an organized system of accepted knowledge that applies in a variety of circumstances to explain a specific set of phenomena; "scientific theories must be falsifiable."

Thermal (energy): Of, relating to, using, producing, or caused by heat.

Thermometer: An instrument for measuring temperature; for example, an instrument with a graduated glass tube and a bulb containing mercury or alcohol that rises in the tube when the temperature increases.

Tide: The alternate rising and falling of the sea due to the attraction of the moon and sun.

Tissue: Any of the distinct types of material of which animals or plants are made, consisting of specialized cells and their products.

Tool: Instruments or utensils that are used to do a particular job.

Transfer: The movement of energy from one location in a system to another system or subsystem.

Troposphere: The lowest region of the atmosphere between Earth's surface and the tropopause, characterized by decreasing temperature with increasing altitude.

Ultraviolet: Electromagnetic radiation having a wavelength just shorter than that of violet light but longer than that of x-rays.

Unicellular: Consisting of a single cell.

Universe: All matter and energy, including Earth, the galaxies, and the contents of intergalactic space, regarded as a whole.

Vapor: Moisture or some other matter visible in the air as mist, clouds, fumes, or smoke.

Variable: Something capable of changing or varying.

Velocity: The speed of something in a given direction.

Versus (vs.): As opposed to or contrasted with.

Vertebrate: Animals having a bony or cartilaginous skeleton with a segmented spinal column and a large brain enclosed in a skull or cranium.

Vibration: An instance of shaking or moving back and forth very rapidly.

Visible light (spectrum): Electromagnetic radiation that can produce a visual sensation.

Volcanic eruption: The sudden occurrence of a violent discharge of steam and volcanic material.

Volcano: A naturally occurring opening in the surface of Earth through which molten, gaseous, and solid material is ejected.

Volume: The loudness of a sound; the size of a three-dimensional space enclosed within or occupied by an object: Symbol V.

Waste: The undigested remainder of food expelled from the body; material, substance, or by product eliminated or discarded as no longer useful.

Glossary

Water: The clear liquid, essential for all plant and animal life, that occurs as rain, snow, and ice, and forms rivers, lakes, and seas.

Wavelength: The distance between successive crests of a wave, especially as a distinctive feature of sound, light, radio waves, etc.

Weather: The state of the atmosphere with regard to temperature, cloudiness, rainfall, wind, and other meteorological conditions.

Weathered: Worn, damaged, or seasoned by exposure to the weather; often used to describe rocks that have been eroded or changed by the action of the weather.

Weathering: Any of the chemical or mechanical processes by which rocks exposed to the weather undergo changes in character and break down.

Weight: The heaviness of somebody or something.

Wind: Air in motion.

White light: Apparently colorless light containing all the wavelengths of the visible spectrum at equal intensity (such as ordinary daylight).

X-ray: An electromagnetic wave of very short wavelength, able to pass through many materials opaque to light.

Year: The time taken by Earth to make one revolution around the sun.

Zoology: The scientific study of the behavior, structure, physiology, classification, and distribution of animals.

This page left intentionally blank.

Science Practice Tutorial

Directions for the Science Practice Tutorial

This Grade 8 CSAP Science Practice Tutorial has multiple-choice and constructed-response questions.

There are several important things to remember as you take the Science CSAP:

- Only No. 2 pencils may be used on any part of the test materials. This includes the front cover.
- During testing, no one is allowed to have electronic communication devices in the testing room.
- Read each multiple-choice question carefully. Think about what is being asked. Then fill in one answer bubble to mark your answer.
- If you do not know the answer to a multiple-choice question, skip it and go on.
- You **may** use the space provided in the test book. You **must s**how all of your work in the space or on the lines provided.
- For constructed-response questions, write your response clearly and neatly on the lines provided.
- If you finish a session early, you may go back and check over your work on that session only. You may be allowed to read but you are **not** allowed to write.

Science Directions—Sample Questions   CSAP Science for Grade 8

Sample Questions

To help you understand how to answer the test questions, look at the sample test questions that follow. They are included to show you what the questions in the test are like and how to mark or write your answers.

Multiple-Choice Sample Question

For this type of question, you will select the answer and fill in the circle next to it. Each multiple-choice question is worth 1 point.

1   Which is the smallest planet in our solar system?

- ● Mercury
- ○ Saturn
- ○ Venus
- ○ Jupiter

For this sample question, the correct answer is the first choice "Mercury"; therefore, the circle next to the first choice is filled in.

Go On

Short or Extended Constructed-Response Sample Question

A constructed-response question may be a short response or extended response type of question. These types of questions have a value of 1–4 score points and you can receive full or partial credit. You should try to answer these questions even if you are not sure of the correct answer. If a part of the answer is correct, you may get a portion of the points.

Be sure to answer every part of the question and use the information provided to help you answer the question.

2 How are people affected when rivers flood?

A flooding river can be devastating to both humans and the environment. Houses are often destroyed and people sometimes drown when a river floods. Damage leads to home evacuation and long-term problems, such as mold and fungus growth in the house. A flooding river can also kill agricultural crops and livestock and cost farmers thousands of dollars. The environmental effect of flooding rivers involves severe erosion of the land and negative altering of the natural landscape.

Go On

Science Practice Tutorial CSAP Science for Grade 8

Question **1** assesses:

Strand: **Science**

Standard 1: Students apply the processes of scientific investigation and design, conduct, communicate about, and evaluate such investigations. **Students know and are able to:**

Benchmark 1.1: Ask questions and state hypotheses that lead to different types of scientific investigations *(for example: experimentation, collecting specimens, constructing models, researching scientific literature).*

Assessment Objective

a: Plan and design a scientific investigation that includes:

- developing a testable question
- researching scientific literature
- stating a hypothesis
- identifying the independent and the dependent variables
- designing a written procedure for a controlled experiment
- using an appropriate observation/measurement technique for data collection
- keeping all other conditions constant

Student Strategies:

For question 1, make sure you are using all of the bulleted points for this Assessment Objective. You may receive up to four score points for completing this constructed-response question. The correct answer must include the following:

- the independent variable
- the dependent variable
- controlled variables
- design features

CSAP Science for Grade 8   Science Practice Tutorial

1   **A science teacher has an idea that students will do better on a test in the morning if they have had breakfast. She would like to test this idea.**

Design an experiment that she can perform to test her idea.

_____

_____

_____

_____

_____

_____

_____

_____

Go On ▶

Analysis: *Constructed-response answers may vary. The teacher could do an experiment with all of her morning classes. Have them take the same test all at the same time. After the test, give them a quick survey on what they had for breakfast. She should score the test and then divide the group into breakfast and no breakfast (the independent variable) and compare the average of the two groups scores (the dependent variable). If the scores are not very different, she can reject her idea. If they are different, she should repeat the experiment a few times to see if the teacher's idea holds up many times.*

Science Practice Tutorial   CSAP Science for Grade 8

Question **2** assesses:

Strand: **Science**

Standard 1: Students apply the processes of scientific investigation and design, conduct, communicate about, and evaluate such investigations. **Students know and are able to:**

Benchmark 1.1: Ask questions and state hypotheses that lead to different types of scientific investigations *(for example: experimentation, collecting specimens, constructing models, researching scientific literature).*

Assessment Objective

b: Identify the independent and dependent variables in a previously conducted scientific investigation on a specific topic.

Student Strategies:

Scientists use controlled experiments to test ideas and hypotheses. A **controlled experiment** seeks to keep all factors, or variables, the same except for the factor that is being tested. All the parts of the experiment that must stay constant are called **controlled variables**. If controlled variables are allowed to change, the results of the experiment may not be valid.

The **independent variable** is the factor that the scientist changes in order to test the hypothesis. For example, if a scientist wants to test whether a certain type of fuel used in cars releases less pollution into the air, the fuel type is the independent variable. To test this, the scientist needs to control the other variables, such as the type of car, the amount of fuel used, etc.

The **dependent variable** is the factor that is measured by the scientist. In the example above, the amount of air pollution released by the cars using different fuels is the dependent variable.

2   A company is performing an experiment to see what kind of cheese children prefer—yellow or white. The company recruits 300 five-year-old and offers them each the two colors of cheese and records which ones the children pick. The company makes sure that the cheeses smell the same, the pieces are the same size, and they are offered the same way every time.

In this experiment, which is the independent variable?

- ○ 300 five-year-old
- ○ color of cheese
- ○ cheese the children choose
- ○ smell of the cheese

Go On

Analysis: *The second choice is correct. The color of the cheese is the variable manipulated by the experimenter. The first choice is incorrect. The children are the sample. The third choice is incorrect. This is the dependent variable. The fourth choice is incorrect. This is a controlled variable because all the cheeses smell the same.*

Science Practice Tutorial   CSAP Science for Grade 8

Question **3** assesses:

Strand: **Science**

Standard 1: Students apply the processes of scientific investigation and design, conduct, communicate about, and evaluate such investigations. **Students know and are able to:**

Benchmark 1.1: Ask questions and state hypotheses that lead to different types of scientific investigations *(for example: experimentation, collecting specimens, constructing models, researching scientific literature).*

Assessment Objective

c: Identify different methods used to investigate scientific questions *(e.g., controlled experiments, collecting specimens, constructing models, researching scientific literature, etc.).*

Student Strategies:

Question 3 is a two-point constructed-response question. The correct answer must include the following:

- The question should be a reasonable, testable, and repeatable question.
- The investigation should reasonably investigate the question.

3 Students traveling on a bus for a field trip notice how the traffic on the highway changes during the trip. Write a valid scientific question that could be asked in this situation and suggest a way the class might investigate the question.

The question:

_____
_____

The investigation:

_____
_____
_____
_____

Go On

Analysis: *Constructed-response answers may vary.*
Question: How does the time of day affect the number of cars on the freeway at milepost 26?
Investigation: The school bus could stop near milepost 26 in the morning and count the cars for 10 minutes. Then the bus could stop again at other times and repeat the observations.

Science Practice Tutorial — CSAP Science for Grade 8

Question assesses:

Strand: **Science**

Standard 1: Students apply the processes of scientific investigation and design, conduct, communicate about, and evaluate such investigations. **Students know and are able to:**

Benchmark 1.2: Use appropriate tools, technologies and metric measurements to gather and organize data and report results.

Assessment Objective

a: Record and report data from a scientific investigation using the appropriate tool and metric units.

Student Strategies:

Here is an example of recorded data from an experiment titled "The Cleanest Lab Table."

Number of Bacterial Colonies Counted

	Alcohol Wash 10 mL	Spray Cleaner 10 mL	Uncleaned Table
Petri dish "alcohol"	4	6	14
Petri dish "spray cleaner"	3	7	15
Petri dish "uncleaned"	4	10	12

Following this experiment, it can be concluded that the lab table that was first cleaned with alcohol wash left the least number of bacteria compared to the table cleaned with the spray cleaner and the uncleaned table.

4 What is the length of the nail shown below?

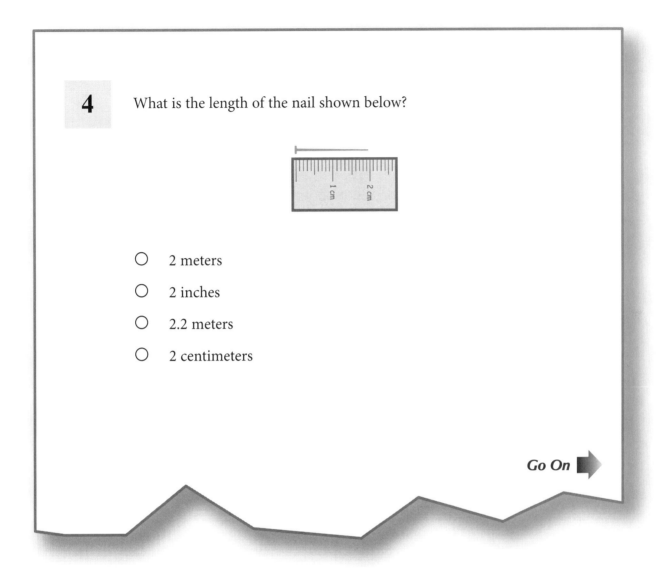

○ 2 meters

○ 2 inches

○ 2.2 meters

○ 2 centimeters

Go On

Analysis: *The fourth choice is correct. Two centimeters is the correct measurement. The first choice is incorrect. Two meters is longer than you are tall. The second choice is incorrect. The scale shown is a part of a meter stick and is in meters, not inches. The third choice is incorrect. The length 2.2 meters is too long.*

Science Practice Tutorial   CSAP Science for Grade 8

Question **5** assesses:

Strand: **Science**

Standard 1: Students apply the processes of scientific investigation and design, conduct, communicate about, and evaluate such investigations. **Students know and are able to:**

Benchmark 1.2: Use appropriate tools, technologies and metric measurements to gather and organize data and report results.

Assessment Objective

b: Describe how different types of technologies are used in scientific investigations *(e.g., telescopes, computers, calculators, seismographs, satellites, microscopes, etc.).*

Student Strategies:

Question 5 is a two-point constructed-response question. The correct answer to this question must include the following:

- two reasonable advantages that a telescope outside Earth's atmosphere and run by computers could have over telescopes on Earth.

5 The Hubble Space Telescope, which is a computerized telescope in orbit around Earth, has led to many advances in our study of the universe.

Name **two** advantages the Hubble Space Telescope has over other telescopes on the surface of Earth.

Advantage 1

_____

_____

Advantage 2

_____

_____

Go On

Analysis: *Constructed-response answers may vary.*
Advantage 1: The Hubble Space Telescope doesn't have to look through the atmosphere, so the view is clearer.
Advantage 2: There is no worry that the Hubble Space Telescope will be affected by the spinning of Earth. The telescope can have a view of Earth and space at all times.

Science Practice Tutorial — CSAP Science for Grade 8

Question **6** assesses:

Strand: **Science**

Standard 1: Students apply the processes of scientific investigation and design, conduct, communicate about, and evaluate such investigations. **Students know and are able to:**

Benchmark 1.2: Use appropriate tools, technologies and metric measurements to gather and organize data and report results.

Assessment Objective

c: Construct and use different types of visual methods *(e.g., data tables, bar and line graphs, diagrams, etc.)* to summarize and present data.

Student Strategies:

Every science student should know by now that different experiments will produce different types of data. Some experiments may result in data that consists of specific measurements of volume, length, or time, while others may include simple comparisons between products.

When an experiment concludes with specific results, a **data table** may work best so that every detail of the data may be shown and understood. In an experiment where a comparison has taken place, a **bar** or **line graph** may work best as a visual representation of the data.

CSAP Science for Grade 8   Science Practice Tutorial

6 Look at the table below about changing temperature of a substance and observations about its phase.

Minutes	Temperature in Degrees Celsius	Phase
1	2°	solid
2	5°	solid and liquid
3	5°	solid and liquid
4	5°	solid and liquid
5	7°	liquid
6	12°	liquid
7	22°	liquid
8	32°	liquid
9	44°	liquid
10	66°	liquid

Which kind of graph would **best** communicate the data in the table?

○ bar graph

○ circle graph

○ histogram

○ line graph

Go On

Analysis: *The fourth choice is correct. A line graph shows data that is happening continuously. The first choice is incorrect. Bar graphs do not show continuous data. The second choice is incorrect. Circle graphs show percentages and proportions. The third choice is incorrect. A histogram is a special bar graph.*

Question **7** assesses:

Strand: **Science**

Standard 1: Students apply the processes of scientific investigation and design, conduct, communicate about, and evaluate such investigations. **Students know and are able to:**

Benchmark 1.3: Interpret and evaluate data in order to formulate a logical conclusion.

Assessment Objective

a: Interpret and evaluate data/observations (*e.g., data tables, bar and line graphs, diagrams, written descriptions, etc.*) to formulate a logical conclusion.

Student Strategies:

Scientific investigation is often about analyzing patterns. Scientists can often learn much about something by finding a pattern and using it to help explain or predict other events.

An example of this can be found in the field of genetics. Scientists have discovered that all living things use DNA as a blueprint for their structures and processes. The pattern and processes of DNA is similar for all living organisms. Scientists are able to use information about DNA from organisms they've studied and apply that knowledge to other organisms they haven't studied.

You can use the process of pattern recognition as well. Try to think about the information you know and the information that is given to you. Look for any patterns and use them to answer questions when you can. Having some knowledge about one system or type of system can often help you answer questions about a different system if you can apply a pattern.

7 Study the data below.

Organism	Number of Chromosomes
Human	46
Rhesus Monkey	42
Cow	60
Dog	78
Cat	38
Horse	64
Mouse	40
Carp	104
Ant	48
Fruit Fly	8
Mosquito	6
Worm	12
Rice	24
Tobacco	48
Potato	48

Which conclusion can be made based on this data?

○ The bigger the organism, the more chromosomes it has.

○ The more complicated the organism, the more chromosomes it has.

○ Plants have fewer chromosomes than animals.

○ No conclusions can be made from this data.

Go On

Analysis: *The fourth choice is correct. There is not enough information here to form a conclusion. The first choice is incorrect. Ants are smaller than mice but have more chromosomes. The second choice is incorrect. It is hard to say which organism is more complicated than the other. Humans are more complicated than dogs, but humans have fewer chromosomes. The third choice is incorrect. Tobacco and potatoes, both of which are plants, have more chromosomes than humans, rhesus monkeys, cats, and mice, all of which are animals.*

Question **8** assesses:

Strand: **Science**

Standard 1: Students apply the processes of scientific investigation and design, conduct, communicate about, and evaluate such investigations. **Students know and are able to:**

Benchmark 1.3: Interpret and evaluate data in order to formulate a logical conclusion.

Assessment Objective

b: Use evidence to state if a hypothesis is supported or not supported.

Student Strategies:

It is important to understand that scientists who criticize theories and discoveries are just as important to the contribution of scientific knowledge as the scientists who do work that supports existing theories, since both types of work result in valid, supportable knowledge.

For this reason, scientists publish their investigations, including the details about the procedures, the data collected, and the conclusions reached, for other scientists to review. As part of this scientific review, other scientists will attempt to duplicate the original investigation in order to obtain the same results. They also review the hypothesis and observations to make sure the scientist used a logical process and conducted the investigation without bias. In this way, scientific investigations and knowledge are tested and we can have confidence in the knowledge obtained.

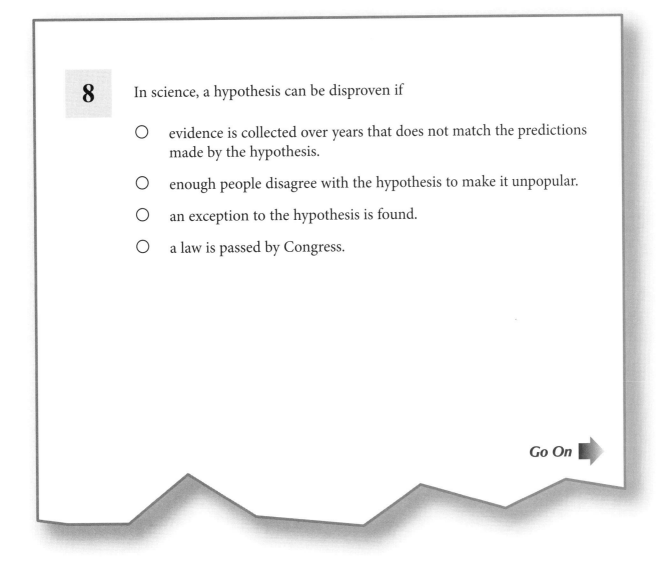

8 In science, a hypothesis can be disproven if

○ evidence is collected over years that does not match the predictions made by the hypothesis.

○ enough people disagree with the hypothesis to make it unpopular.

○ an exception to the hypothesis is found.

○ a law is passed by Congress.

Go On

Analysis: *The first choice is correct. A hypothesis is a strong idea that has been designed to explain observations and evidence collected over a long period of time. Although a hypothesis cannot be proven to be definitely true, it can be disproven if evidence does not support it. The second choice is incorrect. A hypothesis can only be disproven by evidence, not opinion. The third choice is incorrect. It takes a lot of evidence to disprove a hypothesis, not a single exception. The fourth choice is incorrect. Laws do not impact scientific theories or hypotheses.*

Question **9** assesses:

Strand: **Science**

Standard 1: Students apply the processes of scientific investigation and design, conduct, communicate about, and evaluate such investigations. **Students know and are able to:**

Benchmark 1.3: Interpret and evaluate data in order to formulate a logical conclusion.

Assessment Objective

c: Make predictions based on experimental data.

Student Strategies:

Question 9 is a two-point constructed-response question. The correct answer must include the following:

- Two reasonable predictions about what this data shows about the pollutants in the water.

9 Study the experimental data below about water in a mountain stream.

Distance from source	Phosphate Level (mg/L)	Nitrate Level (mg/L)
0 km	0	0
10 km	0	0
20 km	0	0
30 km	0.1	0.94
40 km	0.09	0.91
50 km	0.08	0.89

Make **two** predictions from this data.

Prediction 1

_____

_____

Prediction 2

_____

_____

Go On

Analysis: *Constructed-response answers may vary.*
Prediction 1: There is a home or industry polluting the water between 30 and 50 km from the source of the stream.
Prediction 2: Since the phosphate and nitrate levels decrease after 30 km downstream, there are no other sources of pollution within this distance from the stream.

Question **10** assesses:

Strand: **Science**

Standard 1: Students apply the processes of scientific investigation and design, conduct, communicate about, and evaluate such investigations. **Students know and are able to:**

Benchmark 1.4: Demonstrate that scientific ideas are used to explain previous observations and to predict future events *(for example: plate tectonics and future earthquake activity)*.

Assessment Objective

a: Evaluate collected data/observations and explain the patterns seen in past, current, and future scientific phenomena *(e.g., plate tectonics, future earthquake activity, etc.)*.

Student Strategies:

Question 10 is a two-point constructed-response question. The correct answer must include the following:

- A reasonable time for the next earthquake based on a 10,000-year cycle (+/- a couple of thousand years).

CSAP Science for Grade 8 Science Practice Tutorial

10 Scientists can investigate patterns in the layers of rocks that make up Earth's crust to measure what has happened in the past. For example, the Rio Grande Rift Zone in the San Luis Valley of Southern Colorado shows data of past earthquakes approximately like those listed in the table below.

Quake	Approximate Date
1	45,500 BCE
2	35,500 BCE
3	25,500 BCE
4	15,500 BCE
5	5,500 BCE
6	?

BCE=Before Current Era
Today's date is 2008 CE

When would you predict the next large earthquake (6) will be in the San Luis Valley?

_____

_____

Explain why.

_____

_____

_____

Go On ▶

Analysis: *Constructed-response answers may vary.*
The next earthquake should occur around the year 4,500 CE.
According to the data in the table, the earthquakes seem to occur every 10,000 years. Using simple subtraction, a pattern for the frequency of these earthquakes can be determined.

Question **11** assesses:

Strand: **Science**

Standard 1: Students apply the processes of scientific investigation and design, conduct, communicate about, and evaluate such investigations. **Students know and are able to:**

Benchmark 1.5: Identify and evaluate alternative explanations and procedures.

Assessment Objective

a: Describe other reasonable explanations, using the same independent and dependent variable, for the resulting data or observations from an investigation.

Student Strategies:

Scientists often apply their prior knowledge to scientific questions they have by analyzing patterns and applying these patterns to infer new knowledge. The reason they can do this is because science is based on the assumption that once a scientific rule is discovered and verified, it should hold true for all other things.

This is the basis for scientific investigation. Often, scientists will create a model in order to understand something. The model mimics the processes and characteristics of the scientific question they want to understand. Scientists also use this process to make predictions about processes they may not completely understand.

When you are given a set of data, you should practice looking for patterns. Keep in mind that sometimes there will not be a pattern within the data. However, if data results can be organized into a recognizable pattern, you should be able to use that pattern to predict other results.

11 Investigators observed that in Yellowstone National Park, there are more young aspen trees (dependent variable) growing in the park since wolves (independent variable) started living there in the 1990's. Wolves are one of the top predators in the park.

All of the following are reasonable explanations for this observation **except**

○ wolves reduce the number of elk that eat the young aspen trees.

○ wolves keep the elk moving so they don't have time to eat the aspen trees as much.

○ wolves eat other animals that eat aspen trees.

○ wolves also eat aspen trees and spread their seeds.

Go On

Analysis: *The fourth choice is correct. This is not reasonable since predators (wolves) do not eat plants, including aspen trees. The first choice is incorrect. Wolves reducing the number of elk that eat the young aspen trees is a reasonable explanation for the observation. The second choice is incorrect. Wolves keeping the elk population moving is a reasonable explanation for the observation of the increase of the young aspens. The third choice is incorrect. Wolves eating other animals is a reasonable explanation for the observation.*

Science Practice Tutorial   CSAP Science for Grade 8

Question **12** assesses:

Strand: **Science**

Standard 1: Students apply the processes of scientific investigation and design, conduct, communicate about, and evaluate such investigations. **Students know and are able to:**

Benchmark 1.5: Identify and evaluate alternative explanations and procedures.

Assessment Objective

b: Recognize and/or explain that alternative experimental designs can be used to investigate the same testable question.

Student Strategies:

Question 12 is a four-point constructed-response question. The correct answer must include the following:

- a hypothesis—a reasonable explanation for the observation
- an independent variable—something that is manipulated by the researcher
- a dependent variable—the effect of the test that is measured
- controlled variables—everything that is kept the same

12 A student noticed that one section of the grass in his yard was consistently growing faster than the rest of the grass in the yard. He hypothesized that there might be a broken pipe in the yard and that the grass that was growing faster was getting more water. He wanted to test different ways to make the grass grow faster so he could learn why this was happening. He took a patch of grass and divided it in two. He watered one half of the grass every day. The other half was watered only when it rained. Based on his observations, he concluded that his hypothesis was supported and he started to dig up his yard to find the broken pipe. He couldn't find one.

Design another experiment to test the faster growing grass and explain why that section of the student's yard had grass that grew faster than the rest of the yard.

Hypothesis:

_____

Experiment:

_____

_____

_____

Go On ▶

Analysis: *Constructed-response answers may vary. Hypothesis: The dirt in that section of the yard is better for growing grass than the rest of the yard. Experiment: Collect dirt from different spots in the yard, including the dirt under the grass that is growing fast. Do everything the same to all of the samples, including planting the same kind of grass seed in every dirt sample. If the grass in the special dirt grows faster, then the hypothesis that the dirt is better for grass growth in that section of the yard is supported.*

Question **13** assesses:

Strand: **Science**

Standard 1: Students apply the processes of scientific investigation and design, conduct, communicate about, and evaluate such investigations. **Students know and are able to:**

Benchmark 1.6: Communicate results of their investigations in appropriate ways *(for example: written reports, graphic displays, oral presentations).*

Assessment Objective

a: Recognize that there are several different ways to communicate the results of investigations *(e.g., it is good to keep written reports so that information is preserved over time; oral presentations given to a large group are best when accompanied by a visual presentation; data is best suited for certain types of visual displays—bar graphs, line graphs, tables, etc.)*, and they are each used at different times.

Student Strategies:

It is important for all scientists to keep honest records of their investigations and to report their findings. Scientists must provide all of the data they collected, even if it does not support their hypothesis.

It is also important for other scientists to review the investigation and make criticism. For this reason, a scientific investigation is not considered valid until it has been published in a reputable science magazine. At that point, other scientists can determine if the investigation was conducted correctly and if the data was interpreted in a fair way. Other scientists may also conduct the investigation to see if they obtain similar results.

Scientists who do not keep good records or are not honest about data are not conducting good scientific investigations. Their results cannot be trusted and may create confusion or errors.

13 A class performed an investigation and found some interesting results.

Which of the following could the class do to **best** communicate their results?

○ The class should write a paper for a scientific journal.

○ The class should do a presentation for their community.

○ The class should publish their findings in the school newsletter.

○ all of the above

Go On

Analysis: *The fourth choice is correct. When a scientific investigation is nearly complete, the last step is to publish the results. To communicate their results, the class would reach the greatest audience by publishing in a newsletter and scientific journal as well as present to the community. The first choice is incorrect. Writing a paper is a reasonable option but is not the only solution. The second choice is incorrect. A presentation for their community is a reasonable option but is not the only solution. The third choice is incorrect. Publishing the findings in the school newsletter is a reasonable option but is not the only solution.*

Question **14** assesses:

Strand: **Science**

Standard 2: Physical Science: Students know and understand common properties, forms, and changes in matter and energy. *(Focus: Physics and Chemistry)* **Students know and can demonstrate understanding that:**

Benchmark 2.1: Physical properties of solids, liquids, gases and the plasma state and their changes can be explained using the particulate nature of matter model.

Assessment Objective

a: Describe the particulate model for solid, liquid, gas, and plasma including the arrangement, motion, and energy of the particles *(for example: a lit fluorescent light bulb contains plasma which has widely spaced and highly energetic particles)*.

Student Strategies:

Recall that an object is made up of molecules that are in motion and constantly bumping into each other. Remember that the speed and motion of an object's molecules also determine its physical state. As the kinetic energy of its molecules increases, an object may change from a solid to a liquid, or from a liquid to a gas. In reverse, if the molecules' speed slows down, a gas may change to a liquid, or a liquid may change to a solid. As the energy and speed of the molecules increases, they move more rapidly, bump into each other harder and faster, and bounce farther away from each other.

Visualize water molecules in your mind. You have seen water boil and become steam. The added heat energy increases the speed and distance between the water molecules until the water expands to become a gas. You have also probably seen water vapor condense into a liquid. The water vapor molecules lose energy and move closer together until liquid water forms.

14 **Matter comes in different phases.**

Which of the following has the phases of matter in order from the **least** energy to the **most** energy?

○ solid, liquid, gas, plasma

○ plasma, gas, liquid, solid

○ gas, plasma, solid, liquid

○ liquid, gas, solid, plasma

Go On

Analysis: *The first choice is correct. In a solid, the particles have the least energy, the next are the particles in a liquid, the next in a gas, and the most energy is in the particles of a plasma. The second, third, and fourth choices are incorrect. These choices do not list the phases of matter in order from the least energy to the most energy.*

Question **15** assesses:

Strand: **Science**

Standard 2: Physical Science: Students know and understand common properties, forms, and changes in matter and energy. *(Focus: Physics and Chemistry)* **Students know and can demonstrate understanding that:**

Benchmark 2.1: Physical properties of solids, liquids, gases and the plasma state and their changes can be explained using the particulate nature of matter model.

Assessment Objective

b: Using the kinetic molecular theory, predict how changes in temperature affect the behavior of particles of matter.

Student Strategies:

Kinetic energy is the energy of motion. You should understand that although objects may appear solid, all matter is made of particles that are too small to be seen. Because these particles are bumping into each other very rapidly, they give the object they make up kinetic energy. The kinetic energy of an object increases if energy, such as heat, is transferred to the object. If the object loses energy, the motion of its particles slows down and it loses kinetic energy.

CSAP Science for Grade 8 — Science Practice Tutorial

15 In the two boxes below are particles of a gas and the motion of the particles is greater in one than in the other.

Box 1   Box 2

Which box shows particles at a higher temperature?

○ Box 1

○ Box 2

○ Both are the same temperature.

○ You can't tell from this model.

Go On

Analysis: *The second choice is correct. Temperature is defined in the particle model of matter as the average kinetic energy of the particles—the greater the motion, the higher the temperature. Box 2 shows the particles spaced far apart, which indicates greater motion and higher temperatures. The first choice is incorrect. The particles are close together indicating little motion and lower temperatures. The third choice is incorrect. The two boxes show different motion and different temperatures. The fourth choice is incorrect. Motion is apparent in the model.*

Question **16** assesses:

Strand: **Science**

Standard 2: Physical Science: Students know and understand common properties, forms, and changes in matter and energy. *(Focus: Physics and Chemistry)* **Students know and can demonstrate understanding that:**

Benchmark 2.2: Mixtures of substances can be separated based on their properties *(for example: solubility, boiling points, magnetic properties, densities and specific heat).*

Assessment Objective

a: Explain how to use differences in solubility, boiling points, and magnetic properties to separate mixtures of substances *(for example: filtration can be used to separate mixtures by solubility or physical size).*

Student Strategies:

Mixtures can be separated based on differences in the substances' properties. Some of the most common separation techniques are:

Chromatography—Used for separating different colored dyes. The dyes travel up the chromatography paper different distances. The more soluble dyes move further up than the less soluble ones.

Distillation—Used for separating and collecting a liquid from a solution of a soluble solid. The solution is heated in a flask until the liquid boils. The vapor produced passes into the condenser where it is cooled and condenses to a liquid. The pure liquid (distillate) is collected.

Evaporation—Used to separate a soluble solid from a liquid. If the solution is heated, the liquid evaporates leaving the solid behind.

Fractional Distillation—Used to separate a mixture of liquids. Different liquids boil at different temperatures, so when a mixture of liquids is heated, they boil off and condense at different times.

Filtration—Used to separate an insoluble solid from a liquid. The solid remains in the filter paper and the liquid seeps through the paper into the beaker.

16 Students in a science class have been given a mixture of two white powders that are different substances but don't look different.

How could these **two** substances be separated based on their physical properties?

_____

_____

_____

_____

_____

Go On ▶

Analysis: *Constructed-response answers may vary. These two substances could be separated by putting the mixture of powders in water. Likely, one of the white powders will dissolve and the other will not; for example, sugar and salt are soluble in water. The other powder might not dissolve and sink to the bottom like sand.*

Question **17** assesses:

Strand: **Science**

Standard 2: Physical Science: Students know and understand common properties, forms, and changes in matter and energy. *(Focus: Physics and Chemistry)* **Students know and can demonstrate understanding that:**

Benchmark 2.2: Mixtures of substances can be separated based on their properties *(for example: solubility, boiling points, magnetic properties, densities and specific heat).*

Assessment Objective

b: Apply the concept of density to explain how mixtures of liquids and solids can be separated *(for example: relative densities—sinking and floating).*

Student Strategies:

The **volume** of an object is a measure of how much space it takes up. Volume is measured in units of liters (l) for liquids and cubic centimeters (cm^3) for solids. An object's **density** is a measurement of its mass divided by its volume. Therefore, a dense object will take up less volume than another object that is less dense but has the same mass. For example, iron is denser than wood, so a gram of iron will be smaller, or take up less volume, than a gram of wood.

Relative densities can be determined by whether or not a substance floats or sinks when two substances are mixed. Less dense substances will float, while more dense substances will sink.

17 A tube with five different liquids that have formed five distinct layers is shown below.

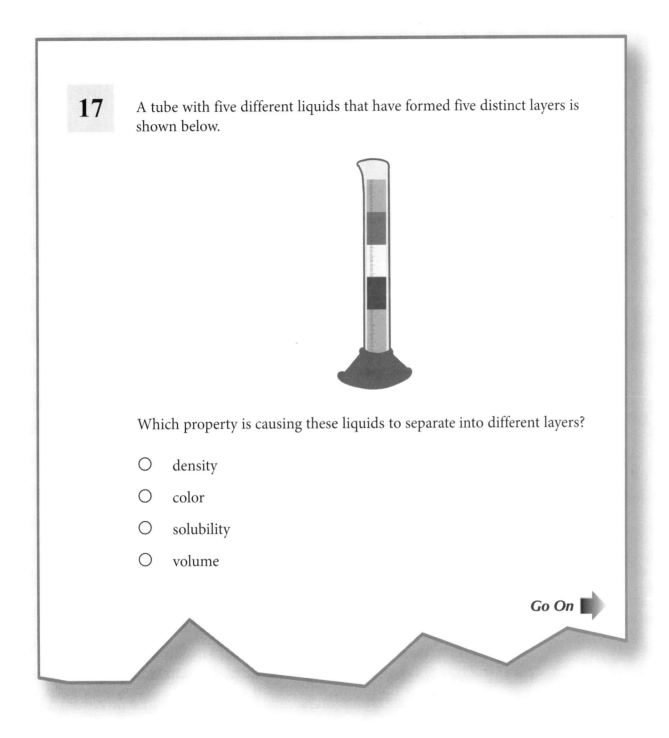

Which property is causing these liquids to separate into different layers?

○ density

○ color

○ solubility

○ volume

Go On

Analysis: *The first choice is correct. Liquids will separate by density if they do not mix. Lighter things float and heavier things sink. The second choice is incorrect. Color doesn't affect sinking and floating. The third choice is incorrect. These liquids do not mix. The fourth choice is incorrect. Volume does not cause liquids to separate into layers.*

Question **18** assesses:

Strand: **Science**

Standard 2: Physical Science: Students know and understand common properties, forms, and changes in matter and energy. *(Focus: Physics and Chemistry)* **Students know and can demonstrate understanding that:**

Benchmark 2.3: Mass is conserved in a chemical or physical change.

Assessment Objective

a: Distinguish between a physical change and a chemical change.

Student Strategies:

Substances can experience many different types of property changes.

A **chemical change** is a change in the chemical composition of a substance. Chemical changes affect a substance at the molecular level and usually create a different substance with different characteristics.

A **physical change** is a change that does not affect a substance chemically, but instead alters a physical characteristic of the substance.

If you are unsure whether a substance has experienced a physical change or a chemical change, it is best to examine the substance before and after the change. Ask yourself if the change has created a substance that acts differently from the original substance. Usually, a physical change leaves a substance looking differently, but you can still recognize the original substance. A chemical change typically creates a new substance that acts very differently from the original substance.

18 Which of the following is an example of a physical change?

- ○ salt dissolving in water
- ○ a nail rusting
- ○ bicarbonate neutralizing acid
- ○ paper burning

Go On

Analysis: *The first choice is correct. A physical change occurs when something is changed and keeps its basic properties. The salt and water keep their basic properties. The second choice is incorrect. The iron in the nail becomes iron oxide which has different properties. The third choice is incorrect. The acid and the bicarbonate react and become a different substance with different properties. The fourth choice is incorrect. The paper is combined with oxygen giving off heat. Ash has different properties than paper.*

Science Practice Tutorial — CSAP Science for Grade 8

Question **19** assesses:

Strand: **Science**

Standard 2: Physical Science: Students know and understand common properties, forms, and changes in matter and energy. *(Focus: Physics and Chemistry)* **Students know and can demonstrate understanding that:**

Benchmark 2.3: Mass is conserved in a chemical or physical change.

Assessment Objective

b: Apply the law of conservation of mass to physical changes *(for example: predict the mass of a substance after a phase change).*

Student Strategies:

The **law of conservation of mass** applies to physical and chemical changes. In physical, or state, changes, the mass of a substance before a change is equal to the mass of the substance after the change. Matter, or mass, is neither created nor destroyed.

For example, if 3 grams of a substance is ground up into a powder, the mass of the powder will still be 3 grams.

19 If 100 grams of water is frozen, what is the mass of the ice?

○ less than 100 grams

○ exactly 100 grams

○ more than 100 grams

○ It is impossible to know.

Go On

Analysis: *The second choice is correct. In a physical change, the total mass of the object changed remains the same. The first, third, and fourth choices are incorrect.*

Question **20** assesses:

Strand: **Science**

Standard 2: Physical Science: Students know and understand common properties, forms, and changes in matter and energy. *(Focus: Physics and Chemistry)* **Students know and can demonstrate understanding that:**

Benchmark 2.3: Mass is conserved in a chemical or physical change.

Assessment Objective

c: Apply the law of conservation of mass to chemical changes *(for example: determine the mass of products given the mass of reactants).*

Student Strategies:

There are many different types of chemical reactions. The most important concept to keep in mind is the conservation of mass. The law of conservation of mass states that mass is neither lost nor destroyed as a result of any regular chemical reaction.

For example, in the equation, $CaF_2 + H_2SO_4 \rightarrow CaSO_4 + 2HF$, the same elements exist on both sides of the equation.

No new elements were created from this chemical reaction, and no elements were eliminated. Even if the reaction produces a gas, the gas will still be shown in the equation. In this equation, calcium fluoride (CaF_2) and sulfuric acid (H_2SO_4) react to create calcium sulfate ($CaSO_4$) and hydrogen fluoride, or hydrofluoric acid (HF). The elements calcium, fluorine, hydrogen, sulfur, and oxygen are found on both sides of the equation in equal amounts. The 2 in front of HF indicates that there are two molecules of hydrogen and two molecules of fluorine present.

20 If you were to put a nail in water in a closed container and weigh it before and after it rusted, how would the mass of the system change?

○ The entire system would lose mass since the nail rusted and lost iron.

○ The entire system would gain mass since the nail gains rust.

○ The entire system would weigh exactly the same since the system was closed.

○ There is no way to know the mass without actually doing the experiment.

Go On

Analysis: *The third choice is correct. In a chemical change, the total mass of the reactants is equal to the total mass of the products. The first choice is incorrect. The system can't lose mass. The second choice is incorrect. The system can't gain mass. The fourth choice is incorrect. We know the mass without actually doing the experiment because of the law of conservation of mass.*

Science Practice Tutorial   CSAP Science for Grade 8

Question **21** assesses:

Strand: **Science**

Standard 2: Physical Science: Students know and understand common properties, forms, and changes in matter and energy. *(Focus: Physics and Chemistry)* **Students know and can demonstrate understanding that:**

Benchmark 2.4: Mass and weight can be distinguished.

Assessment Objective

a: Explain that the mass of an object is the amount of matter *(measured in grams using a balance)* it has and the weight of an object is the force of gravity *(measured in Newtons using a spring scale)* acting on its mass.

Student Strategies:

Substances can differ from each other in a variety of observable ways. You may be able to observe differences in color, size, shape, or texture merely by looking at two objects. Other differences may need to be measured. For example, the mass of an object is the measurement of how much material an object has. You would measure the mass of an object with a balance scale that gives the measurement in grams. The weight of an object is the measure of the force of gravity on the object. You would measure the weight of an object with a scale that gives the measurement in Newtons.

21 The weight of any object near Earth can be measured and reported in which metric unit?

○ Newtons

○ grams

○ pounds

○ ounces

Go On ➡

Analysis: *The first choice is correct. Weight is a measure of the force of gravity on an object. Mass is measured in grams, weight is measured in Newtons. The second choice is incorrect. Mass is measured in grams. The third choice is incorrect. Pounds are a U.S. customary unit of weight. The fourth choice is incorrect. Ounces are a U.S. customary unit of tweight.*

Science Practice Tutorial

Question **22** assesses:

Strand: **Science**

Standard 2: Physical Science: Students know and understand common properties, forms, and changes in matter and energy. *(Focus: Physics and Chemistry)* **Students know and can demonstrate understanding that:**

Benchmark 2.4: Mass and weight can be distinguished.

Assessment Objective

b: Predict how changes in the force of gravity affect the mass and weight of an object *(for example: the mass of an object on the Moon will stay the same but its weight will be less than if the object were on Earth).*

Student Strategies:

While the mass of an object remains the same no matter the location of the object, the weight of an object is a measure of the pull of gravity on the object, and can change depending on the location. The greater the force of gravity, the more weight an object will have, even if its mass remains the same. For example, the force of gravity is about 1/6 as strong on the moon than on Earth. So, an object that has a mass of 1 kilogram and a weight of approximately 10 Newtons on Earth would have a mass of 1 kilogram and a weight of less than 2 Newtons on the moon.

22 The mass of Earth's moon is about one-sixth of the mass of Earth.

How does this affect the force of gravity on an object?

- ○ Things weigh more on the moon than on Earth.
- ○ Things weigh more on Earth than on the moon.
- ○ The mass of a planet does not affect the weight of an object.
- ○ Weight changes but not because of the mass of a planet or moon.

Go On ➡

Analysis: *The second choice is correct. Weight is the measure of the force of gravity on an object. The force of gravity is exactly proportional to the mass of the planet or moon. Earth has more mass, so things will weigh more on Earth than on the moon. The first choice is incorrect. Things will weigh more on Earth since Earth has greater mass than the moon. The third choice is incorrect. The mass of a planet does affect the weight of the object. The fourth choice is incorrect. Weight changes because of the mass of a planet or moon.*

Question **23** assesses:

Strand: **Science**

Standard 2: Physical Science: Students know and understand common properties, forms, and changes in matter and energy. *(Focus: Physics and Chemistry)* **Students know and can demonstrate understanding that:**

Benchmark 2.5: All matter is made up of atoms that are comprised of protons, neutrons and electrons and when a substance is made up of only one type of atom, it is an element.

Assessment Objective

a: Identify that all matter is made up of atoms and that atoms are made of protons, neutrons, and electrons, and describe the location and charge of the parts of an atom.

Student Strategies:

All matter is made up of atoms. An **atom** is the smallest amount of any specific type of matter that can exist. Chemical elements are made up of atoms that have varying amounts of smaller subatomic particles that give them their different characteristics. There are three types of subatomic particles that make up an atom:

Electrons are negatively charged and have the least amount of mass. They move in an orbital cloud at very high speeds around the center of the atom.

Protons are positively charged and have much more mass than electrons. They can be found in the center, or nucleus, of the atom.

Neutrons have no charge and have slightly more mass than protons. They are also found in the nucleus of the atom.

You should be familiar with how the subatomic particles affect the characteristics of an atom. The number of protons and neutrons determines an atom's charge. The number of outer shell electrons determines how an atom reacts with other atoms.

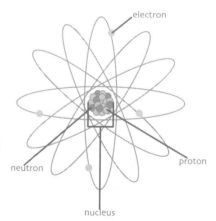

23 Which particle has a positive charge and is found in the nucleus of an atom?

○ electron

○ neutron

○ prion

○ proton

Go On ➤

Analysis: *The fourth choice is correct. A proton is positively charged. The first choice is incorrect. An electron has a negative charge and is located outside the nucleus of the atom. The second choice is incorrect. Neutrons are in the nucleus but carry no charge. The third choice is incorrect. Prions are not found in the nucleus of an atom; they are protein particles that cause diseases like mad cow disease.*

Science Practice Tutorial   CSAP Science for Grade 8

Question **24** assesses:

Strand: **Science**

Standard 2: Physical Science: Students know and understand common properties, forms, and changes in matter and energy. *(Focus: Physics and Chemistry)* **Students know and can demonstrate understanding that:**

Benchmark 2.5: All matter is made up of atoms that are comprised of protons, neutrons and electrons and when a substance is made up of only one type of atom, it is an element.

Assessment Objective

b: Identify that a substance made up of only one type of atom is an element, an atom is the smallest unit of an element that still retains the properties of that element, and different elements have different properties.

Student Strategies:

All matter can be broken down into elements. Elements cannot be changed into simpler substances under normal laboratory conditions. Recall that elements listed on the Periodic Table of Elements are based on an atom's number of protons, neutrons, and electrons.

The smallest amount of any element that still retains the properties of that element is one atom. For example, you may have a bar of the element gold; it is made up of billions of gold atoms, all of which have the same properties.

24 A substance that has only **one** kind of atom in it is called

○ a compound.

○ an element.

○ a mixture.

○ a colloid.

Go On

Analysis: *The second choice is correct. Elements are pure substances made up of only one kind of atom. The first choice is incorrect. Compounds are pure substances composed of two or more kinds of atoms. The third choice is incorrect. Mixtures are composed of different substances with different atoms and molecules. The fourth choice is incorrect. A colloid is a mixture.*

Science Practice Tutorial   CSAP Science for Grade 8

Question **25** assesses:

Strand: **Science**

Standard 2: Physical Science: Students know and understand common properties, forms, and changes in matter and energy. *(Focus: Physics and Chemistry)* **Students know and can demonstrate understanding that:**

Benchmark 2.5: All matter is made up of atoms that are comprised of protons, neutrons and electrons and when a substance is made up of only one type of atom, it is an element.

Assessment Objective

c: Explain that the number of protons in an atom determines what element it is.

Student Strategies:

The **atomic number** of an element is the number of protons present in an atom of that element. Each element has a unique atomic number that determines its placement within the periodic table of elements. For example, the element carbon has 6 protons, so its atomic number is 6. No other element on the periodic table of elements has an atomic number of 6. This is true of every element.

25 In 1869, Dmitri Mendeleev proposed a Periodic Table of the Elements which arranged the pure elements known at that time by their properties. Each element also has an atomic number (e.g., hydrogen is 1 and carbon is 6).

The element's atomic number refers to its number of

- ○ particles.
- ○ electrons.
- ○ protons.
- ○ properties.

Go On ▶

Analysis: *The third choice is correct. An element's atomic number is the number of protons in the nucleus of one of its atoms. The first choice is incorrect. Particles could be neutrons, electrons, protons, or others. The second choice is incorrect. While the number of electrons is often the same as the number of protons, it can vary without changing the element. The fourth choice is incorrect. The number of properties is not related to the element's placement on the periodic table.*

Science Practice Tutorial CSAP Science for Grade 8

Question **26** assesses:

Strand: **Science**

Standard 2: Physical Science: Students know and understand common properties, forms, and changes in matter and energy. *(Focus: Physics and Chemistry)* **Students know and can demonstrate understanding that:**

Benchmark 2.6: When two or more elements are combined a compound is formed which is made up of molecules.

Assessment Objective

a: Explain that two or more atoms may chemically combine to form a molecule, and recognize that a molecule can be represented by a chemical formula that shows the ratio of atoms of each element in the molecule *(for example: H_2 and H_2O are molecules)*.

Student Strategies:

A molecule is a chemically bonded group of atoms that act as a unit, e.g., H_2 (hydrogen gas) or H_2O (water).

The subscript for each element that is present within a molecule determines how many atoms of each element are present within that compound. In the example of water, H_2O, there are 2 hydrogen atoms and 1 oxygen atom present in each molecule of water.

26 In the chemical formula C_2H_5OH, the number of hydrogen atoms in a molecule of ethanol is

- ○ four.
- ○ five.
- ○ six.
- ○ You can't tell from this formula.

Go On

Analysis: *The third choice is correct. A chemical formula shows the exact proportion of atoms in a molecule of a substance: 5 + 1 = 6. The first choice is incorrect. There are five plus one hydrogen atoms in this formula. The second choice is incorrect. Don't forget to add the one atom from the OH group of the formula. The fourth choice is incorrect. The formula is very precise. You are able to tell the number of hydrogen atoms from the information given in the chemical formula.*

Question **27** assesses:

Strand: **Science**

Standard 2: Physical Science: Students know and understand common properties, forms, and changes in matter and energy. *(Focus: Physics and Chemistry)* **Students know and can demonstrate understanding that:**

Benchmark 2.6: When two or more elements are combined a compound is formed which is made up of molecules.

Assessment Objective

b: Describe that two or more elements may chemically combine to form a compound that may have different properties than the elements.

Student Strategies:

A **compound** is a combination of atoms from two or more elements, resulting in a substance that has different properties than the properties of the elements that form it.

An example of a compound is NaCl (table salt), a substance that can be separated into sodium (Na) atoms and chlorine (Cl) atoms in a chemical reaction. This compound exhibits different characteristics than the individual elements sodium and chlorine.

27 If hydrogen and oxygen gases were combined in a container, how could you tell that a new compound has been formed?

○ The new substance has new chemical and physical properties.

○ You can separate them again easily using their original properties.

○ The combination will still burn like hydrogen.

○ The combination will still support burning like oxygen.

Go On

Analysis: *The first choice is correct. When a compound is formed by combining two different substances, the resulting compound has new properties. The second choice is incorrect. If you could separate oxygen and hydrogen easily, it would be a mixture. The third choice is incorrect. A new compound would not retain the original properties of hydrogen. The fourth choice is incorrect. A new compound would not retain the original properties of oxygen.*

Question **28** assesses:

Strand: **Science**

Standard 2: Physical Science: Students know and understand common properties, forms, and changes in matter and energy. *(Focus: Physics and Chemistry)* **Students know and can demonstrate understanding that:**

Benchmark 2.6: When two or more elements are combined a compound is formed which is made up of molecules.

Assessment Objective

c: Explain how mixtures are different than compounds.

Student Strategies:

	Mixtures	Compounds
combination of two or more atoms	X	X
varying composition	X	
constant composition		X
easy to separate	X	
result of physical change	X	
result of chemical change		X
can only be separated by breaking bonds		X

28 Which of the following statements define the difference between a mixture and a compound?

○ A compound is a combination of two or more atoms; a mixture is a combination of only two atoms.

○ A mixture has two or more elements in it; a compound only needs to consist of one element.

○ A compound is the result of a physical change; a mixture is the result of chemical change.

○ A mixture can be easily separated by its physical properties; a compound cannot be easily separated.

Go On

Analysis: *The fourth choice is correct. Mixtures can be separated again using their original physical properties. The first choice is incorrect. Both compounds and mixtures are a combination of two or more atoms. The second choice is incorrect. Both compounds and mixtures must have at least two elements in them. The third choice is incorrect. Compounds are the result of a chemical change.*

Question **29** assesses:

Strand: **Science**

Standard 2: Physical Science: Students know and understand common properties, forms, and changes in matter and energy. *(Focus: Physics and Chemistry)* **Students know and can demonstrate understanding that:**

Benchmark 2.6: When two or more elements are combined a compound is formed which is made up of molecules.

Assessment Objective

d: Identify that the smallest unit of a compound that still retains the properties of that compound is a molecule.

Student Strategies:

A **compound** is a substance of two or more elements combined in fixed ratios by mass. A **molecule** is made of two or more atoms that are combined. All compounds are molecules, but not all molecules are compounds.

Take water for example. Water is a molecule because it is made of atoms that have combined. It is also a compound because it is made of two different types of atoms—hydrogen and oxygen.

Oxygen gas is a molecule because it is made from two atoms of oxygen. It is not a compound because only one element is involved.

29 What is the smallest unit of a compound that still keeps the same properties of the compound?

○ an atom

○ a molecule

○ an element

○ a compound

Go On ➡

Analysis: *The second choice is correct. This is the definition of a molecule—the smallest part of a compound that keeps its properties. The first choice is incorrect. An atom is the smallest part of an element. The third choice is incorrect. An element is composed of atoms. The fourth choice is incorrect. A compound is composed of molecules.*

Question **30** assesses:

Strand: **Science**

Standard 2: Physical Science: Students know and understand common properties, forms, and changes in matter and energy. *(Focus: Physics and Chemistry)* **Students know and can demonstrate understanding that:**

Benchmark 2.7: Quantities *(for example: time, distance, mass, force)* that characterize moving objects and their interactions within a system *(for example: force, speed, velocity, potential energy, kinetic energy)* can be described, measured and calculated.

Assessment Objective

a: Use measurements for objects that are moving in a straight line to relate distance, time, and average speed with words, graphs, and calculations.

Student Strategies:

You can describe the motion of an object in many ways. Motion is the change in position of an object over time. You can describe an object's motion by describing the direction and distance of the change in its position. You can also calculate the speed of the motion by dividing the distance over time: **Average Speed = $\frac{\text{distance}}{\text{time}}$**.

It may be easier to remember that speed, also known as velocity or rate, is represented by the formula: $d = rt$, where d = distance, r = rate (or speed or velocity), and t = time. Notice that with simple division of both sides of the equation, we get: $r = \frac{d}{t}$ to find speed or rate (the same as the equation above) or $t = \frac{d}{r}$ to find the time.

If an object is slowing down or speeding up, it is accelerating. You can calculate acceleration by dividing the change in speed over time:
Acceleration = change in $\frac{\text{speed}}{\text{time}}$.

30 Based on the graphs below of cars moving, which statement is **true**?

Car A

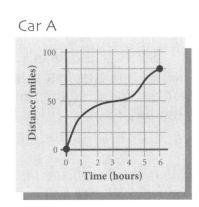

Car B

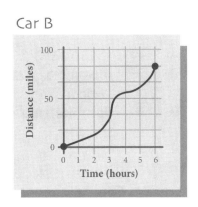

○ Car A is faster than Car B.

○ Car B is faster than Car A.

○ Car A and Car B have the same velocity throughout their paths.

○ Car A and Car B have the same average speed on these trips.

Go On

Analysis: *The fourth choice is correct. These are two graphs of distance vs. time, so they show the speed of the cars. They both reach the destination at the same time even though at several different places on each graph, one car may be moving faster than the other. The first choice is incorrect. The speed varies for both cars on the graphs. Sometimes Car B is faster than Car A. The second choice is incorrect. The speed varies for both cars on the graphs. Sometimes Car A is faster than Car B. The third choice is incorrect. The speeds of both cars vary all along the way.*

Question **31** assesses:

Strand: **Science**

Standard 2: Physical Science: Students know and understand common properties, forms, and changes in matter and energy. *(Focus: Physics and Chemistry)* **Students know and can demonstrate understanding that:**

Benchmark 2.7: Quantities *(for example: time, distance, mass, force)* that characterize moving objects and their interactions within a system *(for example: force, speed, velocity, potential energy, kinetic energy)* can be described, measured and calculated.

Assessment Objective

b: Identify the forces acting on a moving object and explain the effects of changes in the direction and magnitude of forces on the motion of the object.

Student Strategies:

When determining the effect of one or more forces on an object, you should diagram the direction and relative size of the force to help you find the result of the force or forces. For example:

- A. Your brother is applying force to a bicycle (pedals): **bicycle→**.

- B. You apply additional force to the bicycle (push it from the back) as your brother is applying force to the same bicycle: **you→ bicycle→**.
  The bicycle moves forward with the net force (add the forces) that you and your brother are applying to the bicycle. The bicycle moves faster than when he was applying force alone.

- C. You apply an equal but opposite force (push against the bicycle) as your brother applies force (pedals): **bicycle→←you**. When you add the forces together (equal positive and negative forces), they equal zero. The bicycle does not move.

- D. You apply a greater opposite force (push against the bicycle) as your brother applies force (pedals): **bicycle→←←you**. When you add the forces together (his small positive force and your greater negative force), they equal a negative number. The bicycle moves backward.

31 When a person throws a ball, it never goes in a straight line forever as the physicist Sir Isaac Newton predicted.

Which statement explains this effect?

○ Isaac Newton was incorrect about how things travel in space.

○ The push from the person is not the only force on the ball; gravity also pulls the ball down.

○ The force a person uses can only send the ball a little way because the energy runs out.

○ Near Earth, objects usually travel in a curved path on their own without additional forces.

Go On ▶

Analysis: *The second choice is correct. According to Newton's first law of motion, an object in motion continues in a straight path at a constant speed unless acted on by an unbalanced force. This is simpler to see in space rather than near Earth since gravity, wind, and other forces are usually contributing to situations here. There are at least two forces acting on this ball, but there are probably more. The first choice is incorrect. Newton was exactly correct about the motion of objects. That is why his idea is now considered a natural law. The third choice is incorrect. The energy from the person is applied to the ball as kinetic energy and won't change unless something else changes it. The fourth choice is incorrect. All objects move in a straight line unless acted on by another unbalanced force.*

Science Practice Tutorial   CSAP Science for Grade 8

Question **32** assesses:

Strand: **Science**

Standard 2: Physical Science: Students know and understand common properties, forms, and changes in matter and energy. *(Focus: Physics and Chemistry)* **Students know and can demonstrate understanding that:**

Benchmark 2.7: Quantities *(for example: time, distance, mass, force)* that characterize moving objects and their interactions within a system *(for example: force, speed, velocity, potential energy, kinetic energy)* can be described, measured and calculated.

Assessment Objective

c: Compare the relative amount of potential energy *(stored energy)* and kinetic energy *(energy of motion)* of a moving object at different points along its path *(for example: a moving roller coaster has the most potential energy at the top of a hill and the most kinetic energy at the bottom of the hill)*.

Student Strategies:

Energy can exist as potential energy or kinetic energy. **Kinetic energy** is the energy of motion, while **potential energy** is energy due to an object's position that can be released at some future point. You can imagine kinetic energy as a ball rolling down a hill. In this example, the ball's energy is released through its motion. However, if the ball sits at the top of the hill, it has potential energy. The potential energy is due to the position of the ball and the force that gravity exerts on it.

32 The drawing below is of cars on a roller coaster.

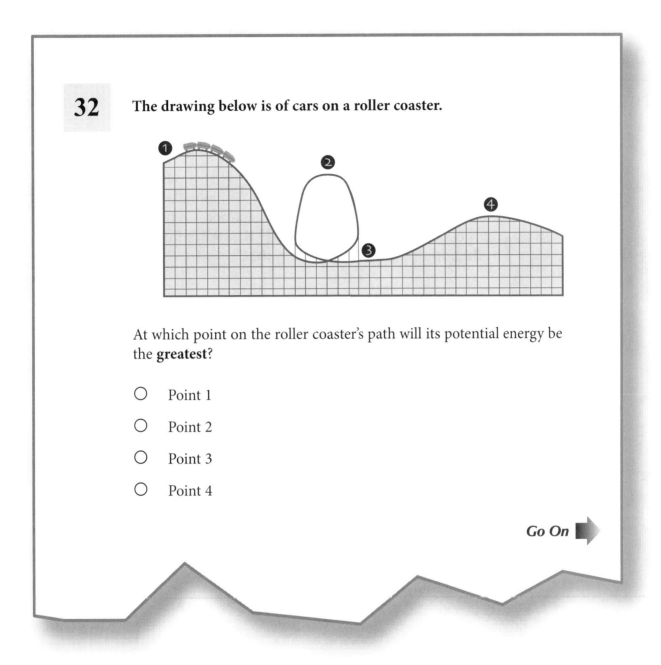

At which point on the roller coaster's path will its potential energy be the **greatest**?

○ Point 1

○ Point 2

○ Point 3

○ Point 4

Go On ▶

Analysis: *The first choice is correct. In a case like this one, potential energy will be the greatest when the car, which is pulled by gravity, is at its highest point. The car is at its highest at Point 1. The second choice is incorrect. Point 1 is higher than Point 2 and has the greatest potential energy. The third choice is incorrect. Point 1 is higher than Point 3. The car probably has the greatest kinetic energy at Point 3. The fourth choice is incorrect. Point 1 is higher than Point 4 and has the greatest potential energy.*

Question **33** assesses:

Strand: **Science**

Standard 2: Physical Science: Students know and understand common properties, forms, and changes in matter and energy. *(Focus: Physics and Chemistry)* **Students know and can demonstrate understanding that:**

Benchmark 2.8: There are different forms of energy and those forms of energy can be transferred and stored *(for example: kinetic, potential)* but total energy is conserved.

Assessment Objective

a: Recognize that energy is the ability to make objects move, and identify that mechanical, sound, thermal, solar, electromagnetic, chemical, and nuclear are some of the forms of energy.

Student Strategies:

Recall that objects are made of molecules that are in motion. The motion, or kinetic energy, of molecules in a substance is also known as its **thermal energy**. Thermal energy can be measured by finding the temperature of the object.

Thermal energy tends to spread out to create equilibrium. That is why when you place a space heater in a room, the heat from the space heater does not stay in front of the heater, but eventually spreads out throughout the room. Thermal energy moves from areas of high thermal energy to areas of low thermal energy. Thinking about it more simply: thermal energy moves from hot areas to colder areas to create equilibrium. Remember that when you are considering these concepts that the terms "hot" and "cold" are relative. Even very cold environments or objects have thermal energy. For example, thermal energy would flow from your refrigerator into your freezer if you made a hole between the two spaces.

33 Which statement about energy is **true**?

○ Energy is the ability to make objects move.

○ Energy is never created or destroyed.

○ Energy can be converted from one form to another.

○ All of the above statements are true.

Go On

Analysis: *The fourth choice is correct. The first three choices about energy are true, so the fourth choice is correct.*

Science Practice Tutorial

CSAP Science for Grade 8

Question **34** assesses:

Strand: **Science**

Standard 2: Physical Science: Students know and understand common properties, forms, and changes in matter and energy. *(Focus: Physics and Chemistry)* **Students know and can demonstrate understanding that:**

Benchmark 2.8: There are different forms of energy and those forms of energy can be transferred and stored *(for example: kinetic, potential)* but total energy is conserved.

Assessment Objective

b: Explain that energy can be transferred *(moved)* from one object to another and transformed *(changed)* from one form to another.

Student Strategies:

Question 34 is a two-point constructed-response question. The correct answer should include the following:

- two energy transformations in this system.

34 Picture a large rock tumbling down a hill. As it falls, it does not keep the same energy.

Describe **two** ways the energy in this system changes.

Energy change 1

_____

_____

Energy change 2

_____

_____

Go On ▶

Analysis: *Constructed-response answers may vary. The falling rock hits objects and makes objects move. The rock makes a sound as it tumbles, creates sparks when it hits objects, releases heat, and the potential energy of the rock is changed into kinetic energy of motion.*

Question **35** assesses:

Strand: **Science**

Standard 2: Physical Science: Students know and understand common properties, forms, and changes in matter and energy. *(Focus: Physics and Chemistry)* **Students know and can demonstrate understanding that:**

Benchmark 2.8: There are different forms of energy and those forms of energy can be transferred and stored *(for example: kinetic, potential)* but total energy is conserved.

Assessment Objective

c: Identify the energy transformations that occur in a specific system.

Student Strategies:

Energy conversions are never completely efficient. When you feel the running engine of an automobile, you feel the heat energy being released as waste instead of being transformed into the motion of the car.

Try to visualize energy transformations that are occurring in any given scenario. For example, coal is burned at an electric power plant and transformed into electricity that travels to a house and is converted into sound and light energy by a television that is plugged into an electrical outlet. Energy is lost at each of these transformations. Energy can take many forms, and energy can also be transformed from or into matter as well. If you can map out the path that energy takes, you can often see the places where energy is being transformed into waste products. Keep in mind that since some energy is always wasted when it is transformed, as more transformations occur, more energy will be lost.

35 When you turn on a light switch in your house, electrical energy can be converted into

- ○ solar energy.
- ○ mechanical energy.
- ○ nuclear energy.
- ○ geothermal energy.

Go On

Analysis: *The second choice is correct. Electrical energy can be converted into mechanical energy. The first choice is incorrect. Electrical energy cannot be converted into solar energy. The third choice is incorrect. Nuclear energy happens when the nucleus of an atom is changed. This doesn't happen in a home electrical system. The fourth choice is incorrect. Geothermal energy is heat energy from Earth.*

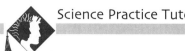

Question **36** assesses:

Strand: **Science**

Standard 2: Physical Science: Students know and understand common properties, forms, and changes in matter and energy. *(Focus: Physics and Chemistry)* **Students know and can demonstrate understanding that:**

Benchmark 2.8: There are different forms of energy and those forms of energy can be transferred and stored *(for example: kinetic, potential)* but total energy is conserved.

Assessment Objective

d: Apply the law of conservation of energy to describe what happens when energy is transferred and/or transformed.

Student Strategies:

The **law of conservation of energy** states that energy cannot be created or destroyed; it can only be changed from one form to another. However, you should remember that energy transfers are not completely efficient. This means that some energy is always lost (given off) as waste energy. "Lost" in this sense means that the energy is not transformed as usable energy. For example, when you plug in an electric light, you are transforming electrical energy into usable light energy. However, some of the electrical energy is also being transformed into heat energy, which you can feel if you touch the light bulb. So when the energy is transformed, the amount of usable energy after the transformation is always less than the total amount of energy that you started with.

36 When the motor on a roller coaster takes the cars all the way to the top of a hill, energy is stored in the cars as potential energy. This energy is converted to motion as the cars descend and is stored again when the cars go uphill.

Why can't the cars continue to travel on the roller coaster forever?

○ Some of the energy is destroyed by friction.

○ Some of the energy makes the passengers scream.

○ Some of the energy is transformed into heat by friction.

○ Some of the energy is changed into mass at the bottom of the hill.

Go On ▶

Analysis: *The third choice is correct. Energy and matter cannot be created or destroyed. The more energy that is lost to heat, the less is available to push the cars up the hill again. The first choice is incorrect. Energy cannot be destroyed by friction. The second choice is incorrect. The energy that makes people scream comes from chemical processes in their bodies. The fourth choice is incorrect. Energy cannot be transformed into mass.*

Question **37** assesses:

Strand: **Science**

Standard 2: Physical Science: Students know and understand common properties, forms, and changes in matter and energy. *(Focus: Physics and Chemistry)* **Students know and can demonstrate understanding that:**

Benchmark 2.9: Electric circuits provide a means of transferring electrical energy when heat, light, sound, magnetic effects and chemical changes are produced.

Assessment Objective

a: Describe the flow of electrons through a circuit.

Student Strategies:

Electrons are able to move in the empty space within and between the atoms of a conductor. The conductor may appear to be solid to our eyes, but since it is composed of atoms, there is plenty of empty space. As each electron moves uniformly through a conductor, it pushes on the one ahead of it, so the electrons move together as a group. Think of a tube full of marbles. If a single marble is inserted into the tube on one side, it will push the marble next to it, which will push the marble next to it and so on, until another marble exits on the other side of the tube. Even though each marble traveled only a short distance, the transfer of motion through the tube is almost instant.

Remember that electrons can flow only when they have the opportunity to move in the space between the atoms of a material. This means that there can be an electric current only when there is a continuous path of conductive material for electrons to travel through.

37 Which statement correctly describes the flow of energy in an electrical circuit?

○ Electrons flow through the wires of a circuit like water flows through a pipe.

○ Electrical energy flows through the space surrounding the metal parts of the circuit.

○ Protons flow through the metal molecules in the wires.

○ None of the above.

Go On

Analysis: *The second choice is correct. Electrical energy moves through the spaces in and around the metal parts of the wires. The first choice is incorrect. Electrons do not actually flow like water. The third choice is incorrect. Protons do not move. The fourth choice is incorrect. The statement in the second choice correctly describes the flow of energy in an electrical current.*

Question **38** assesses:

Strand: **Science**

Standard 2: Physical Science: Students know and understand common properties, forms, and changes in matter and energy. *(Focus: Physics and Chemistry)* **Students know and can demonstrate understanding that:**

Benchmark 2.9: Electric circuits provide a means of transferring electrical energy when heat, light, sound, magnetic effects and chemical changes are produced.

Assessment Objective

b: Identify series circuits and parallel circuits, and compare the two types of circuits.

Student Strategies:

A **series circuit** is a circuit in which resistors are arranged in a chain, so the current has only one path to take.

A **parallel circuit** is a circuit in which the resistors are arranged with their heads connected together, and their tails connected together. The current in a parallel circuit breaks up, with some flowing along each parallel branch and re-combining when the branches meet again.

38 A simple electric circuit including a battery, a switch, and several light bulbs is pictured below.

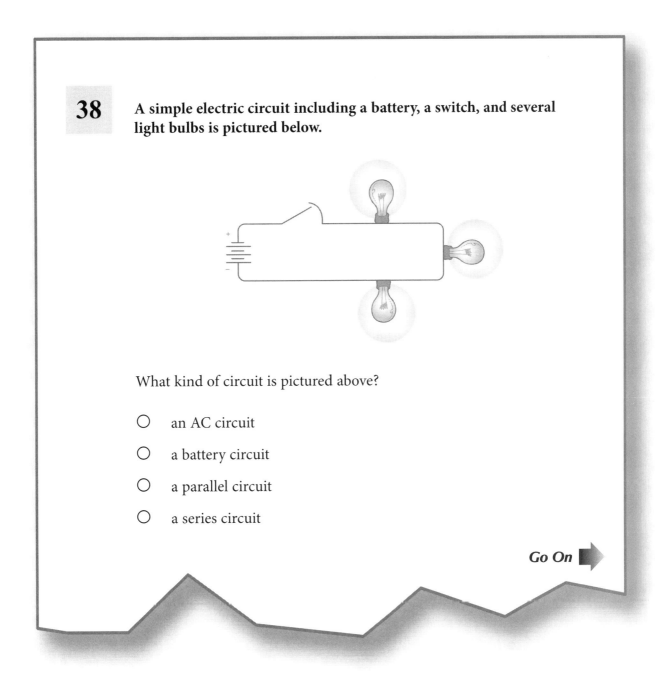

What kind of circuit is pictured above?

○ an AC circuit

○ a battery circuit

○ a parallel circuit

○ a series circuit

Go On

Analysis: *The fourth choice is correct. A circuit that has the flow of energy moving consecutively through its parts is a series circuit. The energy must flow through everything to be a complete circuit. The first choice is incorrect. The battery makes it a DC (direct current) circuit, not an AC (alternating current) circuit. The second choice is incorrect. There is no such thing as a battery circuit. The third choice is incorrect. The lights are not on separate paths, so it could not be a parallel circuit.*

Question **39** assesses:

Strand: **Science**

Standard 2: Physical Science: Students know and understand common properties, forms, and changes in matter and energy. *(Focus: Physics and Chemistry)* **Students know and can demonstrate understanding that:**

Benchmark 2.10: White light is made up of different colors that correspond to different wavelengths.

Assessment Objective

a: Describe that white light is made of different colors of light *(ROYGBIV)*.

Student Strategies:

Visible light waves are the only electromagnetic waves we can see. We see these light waves as the colors of the rainbow: red, orange, yellow, green, blue, indigo, violet. Each color of visible light has a different wavelength; red has the longest wavelength and violet has the shortest wavelength. When light strikes an object, the color that is reflected off the object is the color that we see. For example, a blue car absorbs all of the wavelengths of light with the exception of blue light, which is reflected. When all of the different wavelengths of visible light come together, or are reflected, we see white light.

39 The electromagnetic spectrum includes all of the colors of light that we can see. When the spectrum of light coming from the sun is seen by us on Earth, what is the resulting color of the light?

○ black

○ blue

○ white

○ yellow

Go On ▶

Analysis: *The third choice is correct. When the colors of visible light are mixed, our eyes see it as white light. When white light hits raindrops, it separates into the colors of the spectrum— a rainbow. The first choice is incorrect. If you mix pigments of paint, you get black. Colors of light make white. The second choice is incorrect. Blue is only one of the colors mixed in the spectrum. The fourth choice is incorrect. Yellow is only one of the colors mixed in the spectrum.*

Question **40** assesses:

Strand: **Science**

Standard 2: Physical Science: Students know and understand common properties, forms, and changes in matter and energy. *(Focus: Physics and Chemistry)* **Students know and can demonstrate understanding that:**

Benchmark 2.10: White light is made up of different colors that correspond to different wavelengths.

Assessment Objective

b: Compare the relative wavelengths of different colors of light *(for example: red light has a longer wavelength than blue light)*.

Student Strategies:

Question 40 is a two-point constructed-response question. The answer should place the visible light waves in the proper order from top to bottom:

- 2 points—seven colors in the right place;
- 1 point—five or six colors in the right place;
- 0 points—four or fewer colors in the right place.

40 The colors in the visible spectrum are usually placed in order from the longest wavelength to the shortest wavelength.

Given the colors below, place them in order from longest to shortest.

Blue   Green   Indigo   Orange   Red   Violet   Yellow

Longest _____

_____

_____

_____

_____

_____

Shortest _____

Go On ▶

Analysis: The visible spectrum is usually displayed from longest wavelength to shortest wavelength as **r**ed, **o**range, **y**ellow, **g**reen, **b**lue, **i**ndigo, and **v**iolet (ROYGBIV).

Question **41** assesses:

Strand: **Science**

Standard 3: Life Science: Students know and understand the characteristics and structure of living things, the processes of life, and how living things interact with each other and their environment. *(Focus: Biology—Anatomy, Physiology, Botany, Zoology, Ecology)* **Students know and can demonstrate understanding that:**

Benchmark 3.1: Classification schemes can be used to understand the structure of organisms.

Assessment Objective

a: Identify physical characteristics used to classify vertebrates.

Student Strategies:

Vertebrates are animals that have a backbone. There are five major groups of vertebrates: fish, amphibians, reptiles, birds, and mammals. Each group has characteristics that help scientists classify vertebrates.

Fish: cold-blooded; gills; fins, scales; can be further categorized as jawless fish, cartilaginous fish, or bony fish.

Amphibians: cold-blooded; hatch from eggs laid in water; four limbs; two life stages—gill breathing larvae and lung breathing adults; smooth, moist skin.

Reptiles: cold-blooded; land animals; dry, scaly skin; lay amniotic eggs with leathery shells; can be further categorized into turtles and tortoises with tough shells, meat-eating crocodiles and alligators, or snakes and lizards.

Birds: warm-blooded; body covered with feathers; lay eggs with a hard shell; incubate eggs; beaks; hollow bones.

Mammals: warm-blooded; hair covering body; mammary glands that produce milk to feed their young; care for their young.

41 All vertebrates have one thing in common—they have a backbone. There are several different physical characteristics that help distinguish different classes of vertebrates from one another.

Which characteristic would tell you that an animal is a mammal?

- ○ hair and milk
- ○ feathers and hollow bones
- ○ scales and two heart chambers
- ○ moist skin and three heart chambers

Go On ➡

Analysis: *The first choice is correct. Mammals produce milk to feed their young and have hair covering their bodies. Mammals are also homeotherms (organisms that maintain a constant body temperature) and give birth to live young. The second choice is incorrect. These characteristics describe birds. The third choice is incorrect. These characteristics describe fish. The fourth choice is incorrect. These characteristics describe amphibians.*

Question **42** assesses:

Strand: **Science**

Standard 3: Life Science: Students know and understand the characteristics and structure of living things, the processes of life, and how living things interact with each other and their environment. *(Focus: Biology—Anatomy, Physiology, Botany, Zoology, Ecology)* **Students know and can demonstrate understanding that:**

Benchmark 3.1: Classification schemes can be used to understand the structure of organisms.

Assessment Objective

b: Classify organisms by their physical characteristics *(e.g. using a key, accessing prior knowledge)*.

Student Strategies:

Scientists classify organisms according to their characteristics. Organisms are classified into *Kingdoms, Phylum, Classes, Orders, Family, Genus,* and *Species*. This classification goes in order from general to most specific. The more classification levels that two organisms have in common, the more closely they are related and the more similar they are to each other. For example, although snails and elephants are both in the Animal Kingdom, they are not as closely related as two different species of birds.

Scientists who work on classification are known as **taxonomists**. They examine many characteristics when classifying organisms. These characteristics include size, shape, movement, growth, and reproduction. In addition, scientists often compare the DNA of different organisms to see how similar they are for classification purposes.

42 A student saw a little living thing and wanted to know what it was. It had wings and six legs.

Which field guide should she use to identify what she found?

- ○ Field Guide to Birds
- ○ Field Guide to Spiders and other Arachnids
- ○ Field Guide to Sea Life
- ○ Field Guide to Insects

Go On

Analysis: *The fourth choice is correct. Insects have six legs and sometimes have wings. The first choice is incorrect. Birds have two legs. The second choice is incorrect. Arachnids have eight legs and no wings. The third choice is incorrect. Wings are not typical characteristics of most organisms in the ocean.*

Question **43** assesses:

Strand: **Science**

Standard 3: Life Science: Students know and understand the characteristics and structure of living things, the processes of life, and how living things interact with each other and their environment. *(Focus: Biology—Anatomy, Physiology, Botany, Zoology, Ecology)* **Students know and can demonstrate understanding that:**

Benchmark 3.2: Human body systems have specific functions and interactions *(for example: circulatory and respiratory, muscular and skeletal).*

Assessment Objective

a: Identify organs, organ systems and describe their functions.

Student Strategies:

The major body systems that help the body function include:

Skeletal system: The skeletal system is made up of bones that support the body and help it move.

Circulatory system: The circulatory system is made up of blood vessels that transport blood throughout the body. Blood provides nutrients to the body's cells and collects wastes from them. The heart powers this system.

Muscular system: The muscular system is composed of muscles throughout the body. Muscles are responsible for moving the body, as well as providing the motion for other body organs, such as the heart and lungs, to function.

Digestive system: The digestive system takes in food and breaks it down into nutrients the body can absorb and use. Parts of the digestive system include the mouth, teeth, esophagus, stomach, liver, gallbladder, pancreas, small intestine, and large intestine.

Nervous system: The nervous system processes information about the environment and transmits information and commands throughout the body. The brain and spinal cord are the main components of the nervous system, along with the network of nerve cells that transmit information.

Respiratory system: The respiratory system takes in oxygen and releases carbon dioxide. Gas exchange takes place in the lungs.

Reproductive system: The reproductive system produces male and female gametes that combine to create a fetus during reproduction.

43 Which organ is part of the circulatory system of the human body?

○ veins

○ lymph nodes

○ pharynx

○ ureter

Go On

Analysis: *The first choice is correct. Veins return blood to the heart as a part of the circulatory system. The second choice is incorrect. Lymph nodes are part of the lymphatic and immune systems. The third choice is incorrect. The pharynx is part of the respiratory and digestive systems. The fourth choice is incorrect. The ureter is part of the excretory system.*

Question **44** assesses:

Strand: **Science**

Standard 3: Life Science: Students know and understand the characteristics and structure of living things, the processes of life, and how living things interact with each other and their environment. *(Focus: Biology—Anatomy, Physiology, Botany, Zoology, Ecology)* **Students know and can demonstrate understanding that:**

Benchmark 3.2: Human body systems have specific functions and interactions *(for example: circulatory and respiratory, muscular and skeletal).*

Assessment Objective

b: Explain the interaction of body systems.

Student Strategies:

Question 44 is a two-point constructed-response question. The correct answer should include the following:

- two body systems
- an explanation of how the two body systems work together to accomplish a need.

44 **Human body systems work together to accomplish different functions for the body.**

Name **two** systems that work together in the human body and explain how they work together.

_____

_____

_____

_____

_____

Go On ▶

Analysis: *Constructed-response answers may vary. Two systems in the human body that work together are the circulatory and respiratory systems. The respiratory system brings in oxygen and removes carbon dioxide from the body. The circulatory system moves the oxygen and carbon dioxide around the body and back to the lungs.*

Science Practice Tutorial   CSAP Science for Grade 8

Question **45** assesses:

Strand: **Science**

Standard 3: Life Science: Students know and understand the characteristics and structure of living things, the processes of life, and how living things interact with each other and their environment. *(Focus: Biology—Anatomy, Physiology, Botany, Zoology, Ecology)* **Students know and can demonstrate understanding that:**

Benchmark 3.3: There is a differentiation among levels of organization *(cells, tissues, and organs)* and their roles within the whole organism.

Assessment Objective

a: Identify levels of organization within an organism.

Student Strategies:

Multicellular organisms are all made up of cells at the most basic level. However, similar types of cells can group together and work as **tissues**. Tissues can combine to work together as **organs**. Finally, different organs can combine to create **organ systems**.

Below is an illustration of how this organization works to create parts of the muscular and circulatory system. Heart muscle cells combine to create heart muscle tissue. The heart muscle tissue works together to function as a heart. The heart is the main organ in the circulatory system.

To the right is an illustration of how this organization works to create parts of the muscular and circulatory system. Heart muscle cells combine to create heart muscle tissue. The heart muscle tissue works together to function as a heart. The heart is the main organ in the circulatory system.

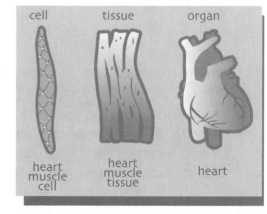

45 One way of organizing how we understand the world is by looking at different levels of organization.

What level within an organism is **smaller** than an organ system?

○ ecosystem

○ biome

○ organ

○ organism

Go On

Analysis: *The third choice is correct. The levels of organization on Earth used in science from largest to smallest are—Biosphere, Biome, Ecosystem, Community, Population, Organism, Organ System, Organ, Tissues, Cells, Organelles, Macromolecules, Molecule, Atom. Organs are within organ systems. The first choice is incorrect. Ecosystems are larger than organ systems. The second choice is incorrect. Biomes are larger than organ systems and ecosystems. The fourth choice is incorrect. An organism is larger than an organ system and organ.*

Question **46** assesses:

Strand: **Science**

Standard 3: Life Science: Students know and understand the characteristics and structure of living things, the processes of life, and how living things interact with each other and their environment. *(Focus: Biology—Anatomy, Physiology, Botany, Zoology, Ecology)* **Students know and can demonstrate understanding that:**

Benchmark 3.4: Multicellular organisms have a variety of ways to get food and other matter to their cells *(for example: digestion, transport of nutrients by circulatory system)*.

Assessment Objective

a: Describe the various processes that food undergoes to be absorbed by an organism's cells.

Student Strategies:

The digestive system prepares nutrients for utilization by body cells through the following processes:

Ingestion: This is the first activity of the digestive system and involves taking food into the mouth.

Mechanical Digestion: This process begins in the mouth with chewing, or mastication, and continues in the stomach where food is mixed. Mechanical digestion is necessary to break down the large pieces of food that are ingested.

Chemical Digestion: This process breaks down the complex molecules of carbohydrates, proteins, and fats into smaller molecules that can be absorbed and used by the cells. Digestive enzymes facilitate and speed up this otherwise slow process.

Movements: After ingestion and mastication, the food particles move into the esophagus by swallowing. Smooth muscles contract to mix the food in the stomach, and continue to contract throughout small segments of the digestive tract to mix the food with enzymes and other fluids and move the food to other parts of the digestive system.

Absorption: This process involves the simple molecules that result from chemical digestion passing through cell membranes of the lining in the small intestine into the blood or lymph capillaries to be used by the cells throughout the body.

46 **Mastication occurs in the mouth to help with digestion.**

How does mastication start the process of digestion?

- ○ It is for tasting food.
- ○ It completes the cooking of food.
- ○ It separates molecules so they can be absorbed.
- ○ It gives the food more surface area so it can be digested.

Go On ➡

Analysis: *The fourth choice is correct. Mastication is the same as chewing. It breaks food up and gives food more surface area so it can quickly and easily be chemically digested in the stomach and intestines. The first choice is incorrect. Tasting is not part of digestion. The second choice is incorrect. Cooking occurs before food is eaten. The third choice is incorrect. The membranes of the intestines separate molecules so they can be absorbed.*

Question **47** assesses:

Strand: **Science**

Standard 3: Life Science: Students know and understand the characteristics and structure of living things, the processes of life, and how living things interact with each other and their environment. *(Focus: Biology—Anatomy, Physiology, Botany, Zoology, Ecology)* **Students know and can demonstrate understanding that:**

Benchmark 3.4: Multicellular organisms have a variety of ways to get food and other matter to their cells *(for example: digestion, transport of nutrients by circulatory system).*

Assessment Objective

b: Identify and compare ways various organisms transport nutrients and wastes *(open and closed circulatory systems, plant vascular systems, etc.).*

Student Strategies:

Vertebrates, and a few invertebrates, have a closed circulatory system. **Closed circulatory systems** are comprised of blood vessels that carry the blood from the heart to the rest of the body.

Mollusks and arthropods have open circulatory systems. **Open circulatory systems** pump blood from the heart into the body cavities, where tissues are surrounded by blood.

Vascular plants transport food, water, and nutrients through their **vascular systems**, which are composed mainly of xylem and phloem. Xylem vessels carry water and minerals and phloem tubes carry food made in photosynthesis.

47 **Nutrients are transported throughout plants by a system of tubes including xylem and phloem cells.**

In which layer of the trunk of a tree would you find these transport tubes?

○ bark

○ cambium

○ wood

○ heartwood

Go On

Analysis: *The second choice is correct. The order from the outside of the stem or trunk to the inside is bark, cambium, wood, and heartwood. The living part is just under the tree bark. The first choice is incorrect. Bark is the outside cork layer of a tree. The third choice is incorrect. Wood is old tissue no longer being used. The fourth choice is incorrect. Heartwood is the oldest part of the wood.*

Question **48** assesses:

Strand: **Science**

Standard 3: Life Science: Students know and understand the characteristics and structure of living things, the processes of life, and how living things interact with each other and their environment. *(Focus: Biology—Anatomy, Physiology, Botany, Zoology, Ecology)* **Students know and can demonstrate understanding that:**

Benchmark 3.4: Multicellular organisms have a variety of ways to get food and other matter to their cells *(for example: digestion, transport of nutrients by circulatory system)*.

Assessment Objective

c: Identify and compare ways various organisms exchange carbon dioxide and oxygen *(stomata, lungs, skin, gills, etc.)* with the environment.

Student Strategies:

There are many different ways organisms exchange gases with the environment.

Stomata, which are tiny openings on the underside of plants' leaves, are used for gas exchange. Carbon dioxide enters through these pores so the plant can conduct photosynthesis; the oxygen product of photosynthesis leaves the plant through these openings. In addition, water vapor is released into the atmosphere through these pores during transpiration.

Lungs, present in air-breathing vertebrates, are gas exchange organs that are primarily responsible for transporting oxygen into the bloodstream and releasing carbon dioxide from the bloodstream into the atmosphere.

Gills, present in underwater animals such as fish, exchange gases through their thin walls of tissue. Gills extract oxygen from water, which is then carried to other parts of the body. Carbon dioxide passes from the body through the gills into the water. Some organisms conduct gas exchange with the environment through their skin. Amphibians have permeable skin that takes in oxygen through tiny blood vessels. Water can also permeate their skin.

48 Exchanging gases with the environment is a critical function for an organism.

Which of the following is a correct pairing of an organism with its gas exchange organ?

○ bird—lung

○ fish—skin

○ turtle—gill

○ insect—lung

Go On

Analysis: *The first choice is correct. Birds have lungs. The second choice is incorrect. Fish have gills, not skin, as a gas exchange organ. The third choice is incorrect. Turtles have lungs; fish have gills. The fourth choice is incorrect. Insects have stoma, not lungs.*

Question **49** assesses:

Strand: **Science**

Standard 3: Life Science: Students know and understand the characteristics and structure of living things, the processes of life, and how living things interact with each other and their environment.
(Focus: Biology—Anatomy, Physiology, Botany, Zoology, Ecology)
Students know and can demonstrate understanding that:

Benchmark 3.5: Photosynthesis and cellular respiration are basic processes of life *(for example: set up a terrarium or aquarium and make changes such as blocking out light).*

Assessment Objective

a: Describe the processes of photosynthesis and cellular respiration.

Student Strategies:

The chemical formula for photosynthesis is: $6CO_2 + 6H_2O + \text{light energy} \rightarrow C_6H_{12}O_6 + 6O_2$. This formula shows what happens when carbon dioxide and water are combined with light energy (often in the form of radiant energy from the sun). The compounds, carbon dioxide and water, are converted by a producer into carbohydrate molecules and oxygen.

A plant uses the carbon dioxide and water molecules, along with the sun's light, to create food. The plant uses the carbohydrates, and releases the oxygen back into the atmosphere.

The radiant energy of the sun is then changed into chemical energy. This chemical energy is the energy source for the plants. Remember that energy can be transferred, transformed, and distributed. Photosynthesis is a great example of how one form of energy (the sun's radiant energy) is transformed into a different form of energy (chemical energy) through a chemical process.

This chemical process is fundamental to the survival of life on Earth. If any one of these things (water, carbon dioxide, or light energy) is missing, photosynthesis cannot occur.

The chemical formula for cellular respiration is: $C_6H_{12}O_6 + 6O_2 \rightarrow 6CO_2 + 6H_2O + ATP$. This formula shows what happens when simple sugars, such as glucose, are combined with oxygen. The compounds are converted into carbon dioxide, water, and usable energy. Both plants and animals use cellular respiration as a means to release energy that the organism can use for life processes.

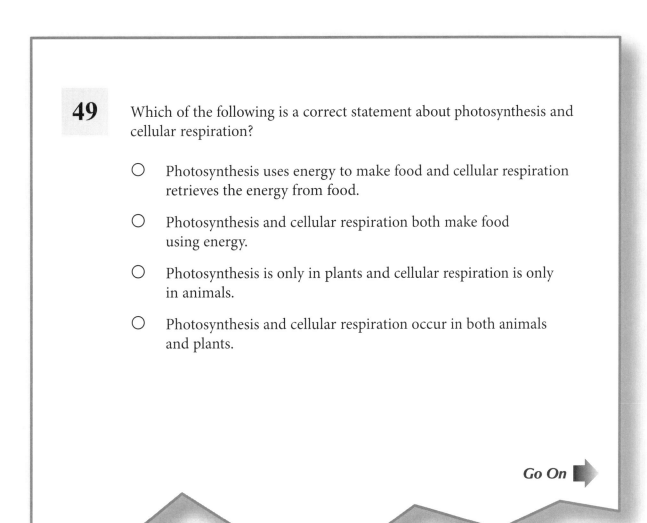

49 Which of the following is a correct statement about photosynthesis and cellular respiration?

○ Photosynthesis uses energy to make food and cellular respiration retrieves the energy from food.

○ Photosynthesis and cellular respiration both make food using energy.

○ Photosynthesis is only in plants and cellular respiration is only in animals.

○ Photosynthesis and cellular respiration occur in both animals and plants.

Go On

Analysis: *The first choice is correct. Photosynthesis uses sunlight as energy and the enzyme chlorophyll to make sugars from carbon dioxide and water. Cellular respiration gets the stored energy back out of the sugars so they can be used by the organism. All organisms utilize cellular respiration. Only organisms with chlorophyll, like plants, can carry on photosynthesis. The second, third, and fourth choices are incorrect.*

Question **50** assesses:

Strand: **Science**

Standard 3: Life Science: Students know and understand the characteristics and structure of living things, the processes of life, and how living things interact with each other and their environment. *(Focus: Biology—Anatomy, Physiology, Botany, Zoology, Ecology)* **Students know and can demonstrate understanding that:**

Benchmark 3.5: Photosynthesis and cellular respiration are basic processes of life *(for example: set up a terrarium or aquarium and make changes such as blocking out light).*

Assessment Objective

b: Describe the relationship between photosynthesis and cellular respiration within plants and between plants and animals *(for example: animals can only do cellular respiration, plants do both).*

Student Strategies:

Question 50 is a two-point constructed-response question. The correct answer will have a completed chart that correctly identifies if each animal or plant performs photosynthesis and/or cellular respiration.

- 2 points—all correct
- 1 point—two correct
- 0 points—none correct

50 Complete the chart below by writing "yes" or "no" in each block for animals and plants depending on whether or not they perform photosynthesis or cellular respiration.

	photosynthesis	cellular respiration
plants		
animals		

Go On

Analysis:

	photosynthesis	cellular respiration
plants	yes	yes
animals	no	yes

Science Practice Tutorial

Question **51** assesses:

Strand: **Science**

Standard 3: Life Science: Students know and understand the characteristics and structure of living things, the processes of life, and how living things interact with each other and their environment. *(Focus: Biology—Anatomy, Physiology, Botany, Zoology, Ecology)* **Students know and can demonstrate understanding that:**

Benchmark 3.6: Different types of cells have basic structures, components and functions *(for example: cell membrane, nucleus, cytoplasm, chloroplast, single-celled organisms in pond water, Elodea, onion cell, human cheek cell).*

Assessment Objective

a: Identify cellular organelles and their functions.

Student Strategies:

To the right are illustrations of the typical plant and animal cells that make up multicellular organisms.

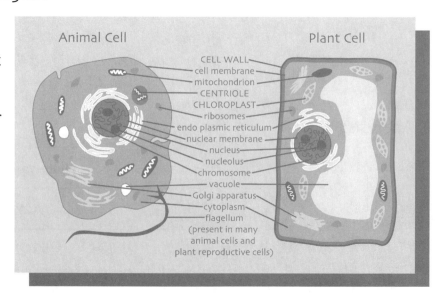

CSAP Science for Grade 8　　　Science Practice Tutorial

51 Label **four** organelles in the plant cell below.

A) _____

B) _____

C) _____

D) _____

Go On

Analysis:
　A) the cell wall
　B) a chloroplast
　C) the nucleus
　D) the cytoplasm

Question **52** assesses:

Strand: **Science**

Standard 3: Life Science: Students know and understand the characteristics and structure of living things, the processes of life, and how living things interact with each other and their environment. *(Focus: Biology—Anatomy, Physiology, Botany, Zoology, Ecology)*
Students know and can demonstrate understanding that:

Benchmark 3.6: Different types of cells have basic structures, components and functions *(for example: cell membrane, nucleus, cytoplasm, chloroplast, single-celled organisms in pond water, Elodea, onion cell, human cheek cell).*

Assessment Objective

b: Differentiate between animal and plant cells and single celled organisms.

Student Strategies:

All living organisms are made up of one or more cells. Most of the organisms found on Earth are single-celled organisms, such as bacteria. However, many more complex organisms, such as animals and plants, are made of many, even millions, of cells that work together.

In multicellular organisms, cells develop and become different from each other in order to perform various jobs. For example, the cells that make up your muscles are very different from the cells that make up your bones.

52 Which organelle is found in plant cells and **not** in animal cells?

- ○ nucleus
- ○ cell wall
- ○ mitochondrion
- ○ cytoplasm

Go On ➡

Analysis: *The second choice is correct. Plant cells are rigid with cell walls. Animal cells do not have cell walls. The first choice is incorrect. Both animal and plant cells have a nucleus. The third choice is incorrect. Both animal and plant cells have a mitochondrion. The fourth choice is incorrect. Both animal and plant cells have cytoplasm.*

Question **53** assesses:

Strand: **Science**

Standard 3: Life Science: Students know and understand the characteristics and structure of living things, the processes of life, and how living things interact with each other and their environment. *(Focus: Biology—Anatomy, Physiology, Botany, Zoology, Ecology)* **Students know and can demonstrate understanding that:**

Benchmark 3.7: There are noncommunicable conditions and communicable diseases *(for example: heart disease and chicken pox).*

Assessment Objective

a: Classify conditions as communicable or noncommunicable and recognize the cause of communicable diseases.

Student Strategies:

A **communicable disease** is a disease that can be transferred from one person to another.

A **noncommunicable disease** is a disease that is not spread from person to person.

53 Diseases can be communicable or noncommunicable based on whether or not you can catch the diseases from another person.

Which of the following is considered noncommunicable?

○ chicken pox

○ the common cold

○ cancer

○ flu

Go On

Analysis: *The third choice is correct. Noncommunicable means you cannot catch the disease from another person. Cancer is not passed among people. The first choice is incorrect. Chicken pox is passed from person to person. The second choice is incorrect. Colds are passed from person to person. The fourth choice is incorrect. The flu is passed from person to person.*

Question **54** assesses:

Strand: **Science**

Standard 3: Life Science: Students know and understand the characteristics and structure of living things, the processes of life, and how living things interact with each other and their environment. *(Focus: Biology—Anatomy, Physiology, Botany, Zoology, Ecology)* **Students know and can demonstrate understanding that:**

Benchmark 3.8: There is a flow of energy and matter in an ecosystem *(for example: as modeled in a food chain, web, pyramid, decomposition).*

Assessment Objective

a: Examine and analyze the flow of energy and matter in a dynamic ecosystem *(e.g., sun to producer to consumer, roles and importance of different organisms).*

Student Strategies:

Energy travels through organisms in a food chain. As energy moves from one level to the next, some energy is lost as heat and other waste products.

Producers are organisms that convert energy from sunlight into sugars that can be used by organisms at each level.

Consumers are organisms that obtain energy from producers or other consumers. Herbivores are organisms that obtain energy directly from eating plants. Herbivores are also known as primary consumers.

Carnivores are organisms that obtain energy indirectly from plants by eating other organisms that have eaten plants. Carnivores are also known as secondary consumers. Carnivores that eat other carnivores may be known as tertiary, or higher, consumers.

Omnivores are organisms that eat both plants and other animals.

Decomposers are organisms, such as bacteria or fungi that break down dead organisms into basic components.

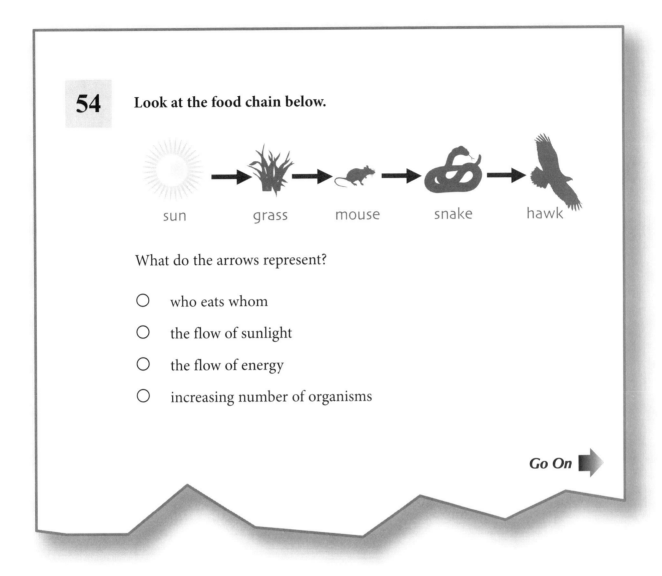

54 Look at the food chain below.

What do the arrows represent?

○ who eats whom

○ the flow of sunlight

○ the flow of energy

○ increasing number of organisms

Go On

Analysis: *The third choice is correct. In a food chain, the arrow represents the direction of the flow of energy. Energy flows along from the lowest forms of life to the top forms of life. The first choice is incorrect. Mice don't eat snakes. The second choice is incorrect. Sunlight only flows into the plant. Other energy flows on along the chain. The fourth choice is incorrect. The number of organisms decreases as you go up the chain.*

Question **55** assesses:

Strand: **Science**

Standard 3: Life Science: Students know and understand the characteristics and structure of living things, the processes of life, and how living things interact with each other and their environment. *(Focus: Biology—Anatomy, Physiology, Botany, Zoology, Ecology)* **Students know and can demonstrate understanding that:**

Benchmark 3.8: There is a flow of energy and matter in an ecosystem *(for example: as modeled in a food chain, web, pyramid, decomposition).*

Assessment Objective

b: Infer the number of organisms or amount of energy available at each level of an energy pyramid.

Student Strategies:

Energy travels through an ecosystem through a series of organisms known as a food chain.

Recall that some energy is always lost when it is transformed. This is true for energy transfers in a food chain as well. When energy is transferred from producers to consumers, typically about 90% of that energy is given off as heat and other waste products, and only 10% is usable by the consumer. This is typical for any energy transfer in the food chain. When an organism is eaten by a carnivore, only 10% of the available energy is useable by the carnivore. As a result, the higher in the food chain an organism is, the fewer of that organism there will be.

55 Look at the food pyramid below.

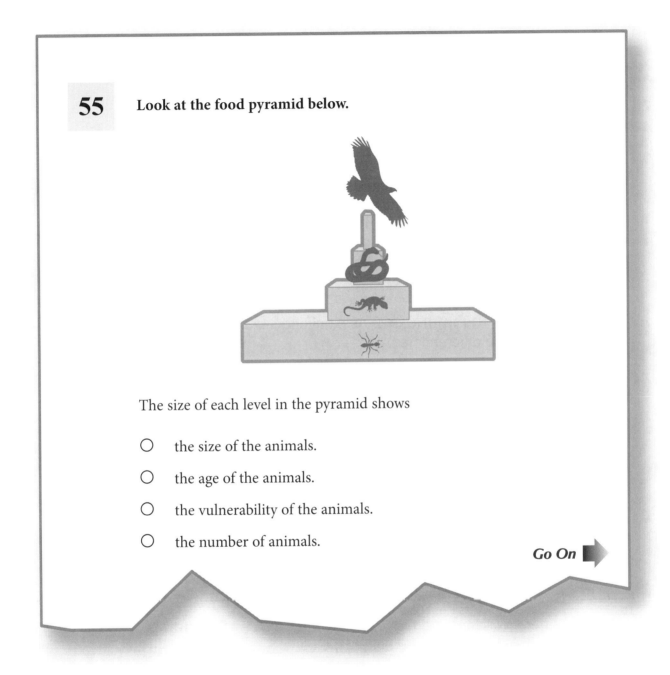

The size of each level in the pyramid shows

○ the size of the animals.

○ the age of the animals.

○ the vulnerability of the animals.

○ the number of animals.

Go On

Analysis: *The fourth choice is correct. In a food pyramid, the decreasing size of the levels shows the amount of biomass at that level. This also typically means the number of organisms at that level. The first choice is incorrect. The size of the animals is not necessarily true in a food pyramid. For example: a wolf is smaller than an elk but higher on the pyramid. The second choice is incorrect. The age of the animals is not important in the animal world. The third choice is incorrect. All of the animals are vulnerable.*

Question **56** assesses:

Strand: **Science**

Standard 3: Life Science: Students know and understand the characteristics and structure of living things, the processes of life, and how living things interact with each other and their environment. *(Focus: Biology—Anatomy, Physiology, Botany, Zoology, Ecology)* **Students know and can demonstrate understanding that:**

Benchmark 3.9: Asexual and sexual cell reproduction/division can be differentiated.

Assessment Objective

a: Differentiate between mitosis and meiosis.

Student Strategies:

The cells of multicellular organisms go through stages called the **cell cycle**. The majority of time a cell spends in the cell cycle is in interphase. In **interphase**, the cell experiences growth and goes about the regular activities of its life. Toward the end of interphase, the cell prepares to divide by making more DNA and creating the materials needed for division, or **mitosis**. The phases of mitosis are shown to the right:

At the end of mitosis, the cell contains two nuclei with identical DNA. At this point, the cell undergoes **cytokinesis**, the process in which the cell itself splits in two, with a nucleus in each new cell.

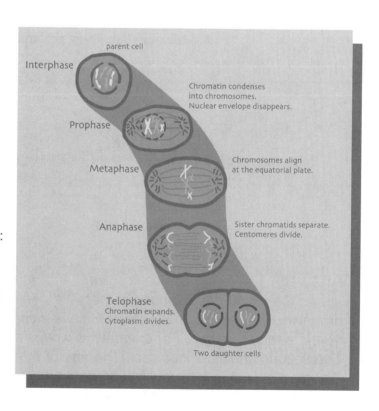

56 What is the difference between mitosis and meiosis?

○ Meiosis results in two cells identical to the parent cell.

○ Mitosis results in two cells identical to the parent cell.

○ Meiosis happens in every cell in the body.

○ Mitosis only occurs in sex organs.

Go On ➡

Analysis: *The second choice is correct. Mitosis is a body process that reproduces identical cells to the parent. Meiosis results in haploid cells and occurs in sex organs. The first, third, and fourth choices are incorrect.*

Question **57** assesses:

Strand: **Science**

Standard 3: Life Science: Students know and understand the characteristics and structure of living things, the processes of life, and how living things interact with each other and their environment. *(Focus: Biology—Anatomy, Physiology, Botany, Zoology, Ecology)* **Students know and can demonstrate understanding that:**

Benchmark 3.9: Asexual and sexual cell reproduction/division can be differentiated.

Assessment Objective

b: Relate the number of chromosomes to the final product of mitosis or meiosis.

Student Strategies:

Meiosis is the process in which a diploid eukaryotic cell divides into four haploid cells, or gametes. Meiosis is essential for sexual reproduction. The phases of meiosis are shown to the right.

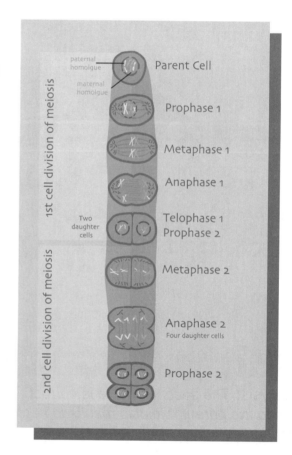

57 Which statement is **true** for cell division?

○ Both mitosis and meiosis begin with diploid cells.

○ Both mitosis and meiosis result in diploid cells.

○ Cells dividing through meiosis result in diploid cells.

○ Cells dividing through mitosis result in haploid cells.

Go On

Analysis: *The first choice is correct. Mitosis is a body process that starts and ends with diploid cells. Meiosis begins with diploid cells and results in haploid cells. The second, third, and fourth choices are incorrect.*

Question **58** assesses:

Strand: **Science**

Standard 3: Life Science: Students know and understand the characteristics and structure of living things, the processes of life, and how living things interact with each other and their environment. *(Focus: Biology—Anatomy, Physiology, Botany, Zoology, Ecology)* **Students know and can demonstrate understanding that:**

Benchmark 3.10: Chromosomes and genes play a role in heredity *(for example: genes control traits, while chromosomes are made up of many genes).*

Assessment Objective

a: Describe the relationship between chromosomes, genes and traits and their role in heredity.

Student Strategies:

DNA (deoxyribonucleic acid) is a molecule that is the building unit of genetic material.

DNA molecules combine with each other to form a very long chain, which is called a **chromosome**. A single human chromosome can consist of millions of DNA molecules linked in a chain.

Within chromosomes, smaller areas can be defined that are made up of a few thousand DNA molecules. These are called genes, and are responsible for encoding the genetic characteristics of living organisms. Genes are responsible for the different traits we see in organisms.

58 The complicated instructions and patterns needed to produce and control the appearance and function of organisms are found on the

○ nuclear membrane.

○ chromosomes.

○ nucleolus.

○ genes.

Go On

Analysis: *The fourth choice is correct. Genes control hereditary traits, which include the appearance and function of an organism. The first choice is incorrect. The nuclear membrane contains the nuclear region where the chromosomes and genes exist. The second choice is incorrect. Chromosomes are the body in the cell on which the genes reside. The third choice is incorrect. The nucleolus is in the nucleus and has no relation to genetic traits.*

Science Practice Tutorial   CSAP Science for Grade 8

Question **59** assesses:

Strand: **Science**

Standard 3: Life Science: Students know and understand the characteristics and structure of living things, the processes of life, and how living things interact with each other and their environment. *(Focus: Biology—Anatomy, Physiology, Botany, Zoology, Ecology)*
Students know and can demonstrate understanding that:

Benchmark 3.10: Chromosomes and genes play a role in heredity *(for example: genes control traits, while chromosomes are made up of many genes).*

Assessment Objective

b: Infer the traits of the offspring based on the genes of the parents *(including dominant, recessive traits and use of Punnet square diagrams).*

Student Strategies:

Sexual reproduction creates new combinations of traits in the offspring of organisms because each parent contributes different genes that are passed on. You should be familiar with Punnett squares and how they work.

You should first know whether a gene is dominant or recessive. **Dominant** genes are expressed even if there is only one dominant gene present. They are usually indicated by a capital letter. **Recessive** genes are the unexpressed genes, usually indicated by a lowercase letter. For example, if B represents the gene that gives a dog brown fur and b represents the gene that gives a dog white fur, then a dog with one dominant gene and one recessive gene would be designated by Bb and would have brown fur. A BB dog would have both dominant genes and be brown, while a bb dog would have only recessive genes and would be white.

You can fill out a Punnett square to determine the probability of different genes being passed on to offspring. If you had a dog with brown fur that had both dominant genes for brown fur mate with a dog that had both recessive genes for white fur, you might make a table like the table to the right.

		Dominant brown-fur dog	
		B	B
Recessive white-fur dog	b	Bb	Bb
	b	Bb	Bb

The dominant dog would pass along either one B gene or another B gene. The recessive dog would pass one b gene or another b gene. In this case, all offspring would be Bb, with brown fur.

59 Consider the Punnet square below dealing with gray seeds (G) and white seeds (g).

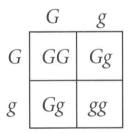

What percentage of the offspring will be gray?

○ 25%

○ 50%

○ 75%

○ 100%

Analysis: *The third choice is correct. In a case like this, gray is dominant to the white color in seeds. Both GG and Gg genotypes would produce gray seeds, and three out of four are gray. The first choice is incorrect. The genotype gg is white, one out of four of the offspring will be white. The second choice is incorrect. Gg and GG are both gray. The fourth choice is incorrect. The genotype gg is white, GG is gray, and Gg is gray.*

Question **60** assesses:

Strand: **Science**

Standard 3: Life Science: Students know and understand the characteristics and structure of living things, the processes of life, and how living things interact with each other and their environment. *(Focus: Biology—Anatomy, Physiology, Botany, Zoology, Ecology)* **Students know and can demonstrate understanding that:**

Benchmark 3.11: Changes in environmental conditions can affect the survival of individual organisms, populations, and entire species.

Assessment Objective

a: Describe several environmental factors that could limit the size of an organism's population.

Student Strategies:

Question 60 is a two-point constructed-response question. The correct answer should include:

- two factors that could limit the size of a population.

60 **Sometimes things change in an environment that affect the number of organisms in a population.**

Name **two** factors that would limit the size of a population.

_____

_____

_____

_____

Go On ➡

Analysis: *Constructed-response answers may vary. The list of factors could include: food, nesting sites, shelter sites, availability of mates, etc.*

Question **61** assesses:

Strand: **Science**

Standard 3: Life Science: Students know and understand the characteristics and structure of living things, the processes of life, and how living things interact with each other and their environment. *(Focus: Biology—Anatomy, Physiology, Botany, Zoology, Ecology)* **Students know and can demonstrate understanding that:**

Benchmark 3.11: Changes in environmental conditions can affect the survival of individual organisms, populations, and entire species.

Assessment Objective

b: Describe the impact of humans on the environment and how that affects the survival of populations and entire species.

Student Strategies:

Question 61 is a two-point constructed-response question. The correct answer should include:

- two ways humans can impact the environment.

61 By living in areas where other organisms inhabit, humans change the environmental conditions for those organisms.

Describe **two** ways humans impact the environment.

_____

_____

_____

_____

_____

Go On ➡

Analysis: *Constructed-response answers may vary. The changes listed could include: pollution, habitat destruction, cutting down trees, planting different plants, etc.*

Question **62** assesses:

Strand: **Science**

Standard 3: Life Science: Students know and understand the characteristics and structure of living things, the processes of life, and how living things interact with each other and their environment. *(Focus: Biology—Anatomy, Physiology, Botany, Zoology, Ecology)*
Students know and can demonstrate understanding that:

Benchmark 3.11: Changes in environmental conditions can affect the survival of individual organisms, populations, and entire species.

Assessment Objective

c: Describe how organisms change in response to environmental factors.

Student Strategies:

Although new physical characteristics or behaviors that organisms exhibit are usually random in origin, they could be helpful or harmful to the organism's ability to survive and reproduce. If the adaptation is helpful, it will increase in occurrence because it will allow those organisms to live and reproduce, passing on their characteristics to new offspring.

The new characteristics that develop can be either a change in a specific structure of an organism, such as a change in a bird's beak, or an animal's claw, or the change can also take the form of a behavior, such as a mating ritual or nesting ability. It may also take the form of a difference in physiology, or body type.

When thinking about whether a change is adaptive (helpful) or non-adaptive (harmful), it is wise to consider the organism's environment. Most organisms develop specific characteristics or behaviors that help them adapt to a specific environment, or niche. By becoming better at surviving in this specific environment, they are able to out-compete other organisms' lack of special skills or structures.

Specialization can make them successful in their environment. For example, polar bears are white. Since they live in snowy environments, their white fur helps them hide from potential prey.

62 When changes in the environment occur, the organisms that live there must respond.

Which of the following is a likely response to a drastic decrease in the size of a habitat?

○ have more offspring

○ breed more often

○ leave for a better habitat

○ reconstruct the habitat

Go On ▶

Analysis: *The third choice is correct. Organisms will have to respond in a way that leads to their survival, which may include leaving the area or having fewer offspring. The first choice is incorrect. Having more offspring would result in a catastrophe since there is less living area. The second choice is incorrect. Breeding more often would also result in a catastrophe since there is less living area. The fourth choice is incorrect. Reconstructing the habitat is beyond the ability of many organisms.*

Question **63** assesses:

Strand: **Science**

Standard 3: Life Science: Students know and understand the characteristics and structure of living things, the processes of life, and how living things interact with each other and their environment. *(Focus: Biology—Anatomy, Physiology, Botany, Zoology, Ecology)* **Students know and can demonstrate understanding that:**

Benchmark 3.12: Changes or constancy in groups of organisms over geologic time can be revealed through evidence.

Assessment Objective

a: Compare and contrast evidence of past life from different epochs to existing organisms.

Student Strategies:

Fossils are created when organisms die and their bodies are preserved in rock, mud, or any medium that hardens and survives over thousands or millions of years. Although the occurrence of fossils varies throughout history, there are still many organisms from long ago that scientists can study today because they became fossils.

63 **Fossils frequently resemble organisms that are still living today.**

What does this say about organisms on Earth?

○ Life on Earth has been around a very long time since fossils take so long to form.

○ Life forms change very slowly.

○ Some habitats on Earth have changed very little over geologic time.

○ All of the above are correct.

Go On

Analysis: *The fourth choice is correct. It takes centuries for fossils to form; some habitats have changed very little over geologic time; and frequently we find great similarities between life forms from today and those that were fossilized.*

Science Practice Tutorial   CSAP Science for Grade 8

Question assesses:

Strand: **Science**

Standard 3: Life Science: Students know and understand the characteristics and structure of living things, the processes of life, and how living things interact with each other and their environment. *(Focus: Biology—Anatomy, Physiology, Botany, Zoology, Ecology)* **Students know and can demonstrate understanding that:**

Benchmark 3.13: Individual organisms with certain traits are more likely than others to survive and have offspring.

Assessment Objective

a: Evaluate the potential of an organism with specific traits to survive and reproduce in an environment.

Student Strategies:

The ultimate goal for behavior or characteristics of an organism or species is to survive and reproduce. When you must evaluate a characteristic or change, try to think about how that change might help or hurt an organism's ability to survive or create offspring.

Over many generations, characteristics or behaviors that help an organism survive will typically become more common. Characteristics that hurt an organism's chances of surviving or reproducing will eventually disappear as organisms with these characteristics do not survive long enough to reproduce. Most new characteristics or behaviors begin as random events. It is only after several generations of reproduction that some of these traits become more common as they have caused individuals that have the traits to be more successful than other organisms.

64 In 1859, an English scientist named Charles Darwin published decades of research on animals and proposed the theory of natural selection.

Which statement is **true** according to this theory?

○   People came from apes.

○   Animals can change their bodies to adapt to changing conditions.

○   The strongest organisms live to have offspring.

○   People cannot manipulate which animal traits are inherited.

Go On

Analysis: *The third choice is correct. Darwin's "The Origin of the Species" proposed the theory of natural selection which is often referred to as "survival of the fittest." The first choice is incorrect. Darwin proposed that humans and apes might have a common ancestor a very long time ago in a different book, "The Descent of Man." The second choice is incorrect. Once an organism has developed, it can no longer adapt its characteristics, only survive to reproduce or not. The fourth choice is incorrect. People manipulate traits of dogs and other animals through selective breeding.*

Science Practice Tutorial   CSAP Science for Grade 8

Question **65** assesses:

Strand: **Science**

Standard 4: Earth and Space Science: Students know and understand the processes and interactions of Earth's systems and the structure and dynamics of Earth and other objects in space. *(Focus: Geology, Meteorology, Astronomy, Oceanography)*
Students know and can demonstrate understanding that:

Benchmark 4.1: Inter-relationships exist between minerals, rocks and soils.

Assessment Objective

a: Understand the three types of rocks *(igneous, sedimentary, metamorphic)* and the processes that formed them through the rock cycle.

Student Strategies:

Rock Cycle

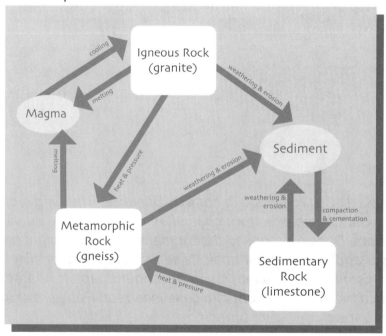

186 Student Self-Study Workbook   Copying is Prohibited   © Englefield & Associates, Inc.

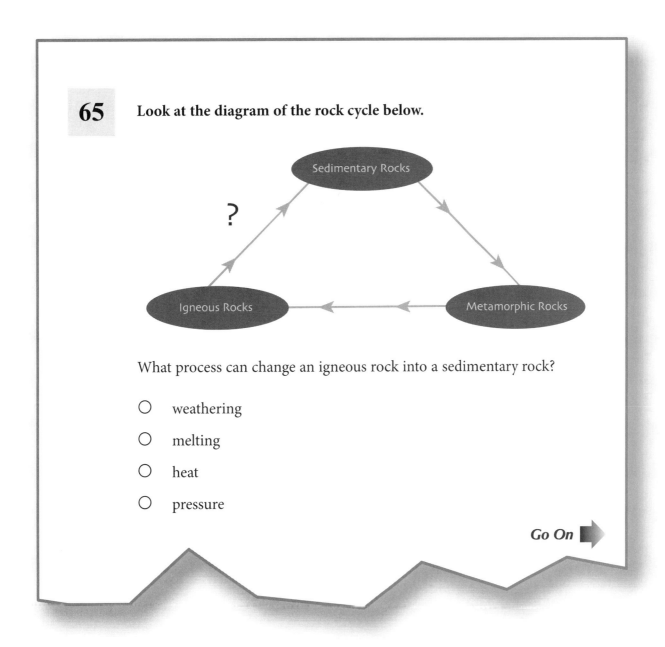

65 Look at the diagram of the rock cycle below.

What process can change an igneous rock into a sedimentary rock?

○ weathering

○ melting

○ heat

○ pressure

Go On

Analysis: *The first choice is correct. To make a sedimentary rock, weathering and deposition must occur. The second choice is incorrect. Melting would make an igneous rock. The third choice is incorrect. Heat helps make both metamorphic and igneous rocks depending on the temperature. The fourth choice is incorrect. Pressure helps make metamorphic rocks.*

Science Practice Tutorial — CSAP Science for Grade 8

Question **66** assesses:

Strand: **Science**

Standard 4: Earth and Space Science: Students know and understand the processes and interactions of Earth's systems and the structure and dynamics of Earth and other objects in space. *(Focus: Geology, Meteorology, Astronomy, Oceanography)*
Students know and can demonstrate understanding that:

Benchmark 4.1: Inter-relationships exist between minerals, rocks and soils.

Assessment Objective

b: Understand the composition and relationships of rocks, minerals, and soil formation.

Student Strategies:

Soil formation takes a very long time. Soil is formed from the weathering of rocks and minerals. As rock breaks down into smaller pieces through weathering, it is mixed with moss and organic matter. Plants can help the development of soil by attracting animals, which, in turn, contribute to organic matter: when animals die, they decay and make rich soil. Over time, this creates a thin layer of soil. This continues until the soil is fully formed, after which it supports many different plants and the process continues.

66 Rocks are groups of minerals that have formed together into masses.

What characterizes the different minerals?

○ the area where they formed

○ different chemical composition

○ how they were formed

○ their value

Go On ▶

Analysis: *The second choice is correct. Minerals are classified by chemical composition. The first choice is incorrect. Minerals of like kinds are found all over the planet. The third choice is incorrect. Rocks, not minerals, are classified by how they are formed. The fourth choice is incorrect. The value of the mineral is not a scientific concept; it is economic.*

Question **67** assesses:

Strand: **Science**

Standard 4: Earth and Space Science: Students know and understand the processes and interactions of Earth's systems and the structure and dynamics of Earth and other objects in space. *(Focus: Geology, Meteorology, Astronomy, Oceanography)*
Students know and can demonstrate understanding that:

Benchmark 4.2: Humans use renewable and nonrenewable resources *(for example: forests and fossil fuels)*.

Assessment Objective

a: Understand the differences between renewable and nonrenewable energy resources.

Student Strategies:

Energy is obtained from either renewable or nonrenewable resources. A **renewable resource** is one that does not run out. Some renewable resources include solar energy, wind power, nuclear energy, hydroelectric power, and geothermal energy. Although renewable energy sources provide an endless supply of energy, there are drawbacks. Some renewable energy resources release pollution, while others are costly or less efficient than nonrenewable resources.

Most of the energy used by society right now comes from **nonrenewable sources**. Humans rely mostly on fossil fuels, such as oil, coal, and natural gas. Fossil fuels come from living organisms that died millions of years ago, leaving remains that have changed with time and pressure. Although using fossil fuels is fairly cheap, they release large amounts of carbon dioxide and other pollutants into the air. Since supplies of nonrenewable resources are limited, it is important for humans to find other sources of energy to replace them when they are gone.

67 Which of the following is a nonrenewable source of energy?

○ solar

○ oil

○ hydroelectric

○ geothermal

Go On

Analysis: *The second choice is correct. Oil is a nonrenewable resource. A nonrenewable source of energy is one that takes so long to make that it cannot be replaced in our lifetime. The first choice is incorrect. The sun is a source of energy that is seemingly inexhaustible as long as we don't block the sky. The third choice is incorrect. When water moves through a generator, it makes electricity. Hydroelectric energy is renewable as long as we have gravity and can find water. The fourth choice is incorrect. Geothermal energy is a renewable resource that captures the power from steam from the earth to generate electricity.*

Science Practice Tutorial CSAP Science for Grade 8

Question **68** assesses:

Strand: **Science**

Standard 4: Earth and Space Science: Students know and understand the processes and interactions of Earth's systems and the structure and dynamics of Earth and other objects in space. *(Focus: Geology, Meteorology, Astronomy, Oceanography)*
Students know and can demonstrate understanding that:

Benchmark 4.2: Humans use renewable and nonrenewable resources *(for example: forests and fossil fuels)*.

Assessment Objective

b: Predict the advantages and disadvantages of using both types of energy resources *(renewable and nonrenewable)* and their sustainability.

Student Strategies:

Question 68 is a two-point constructed-response question. The correct answer should include:

- one advantage of using coal as a source of energy
- one disadvantage of using coal as a source of energy

68 Name an **advantage** and a **disadvantage** of using coal as a source of energy.

Advantage:

_____

_____

Disadvantage:

_____

_____

Go On ▶

Analysis: *Constructed-response answers may vary. There are advantages and disadvantages to using coal as a source of energy. Some advantages could include: our familiarity with coal, or that it is relatively cheap to use with current technology. Some disadvantages could include: it is a dirty, polluting source of energy, or that it is nonrenewable and we can't get more.*

Question **69** assesses:

Strand: **Science**

Standard 4: Earth and Space Science: Students know and understand the processes and interactions of Earth's systems and the structure and dynamics of Earth and other objects in space. *(Focus: Geology, Meteorology, Astronomy, Oceanography)*
Students know and can demonstrate understanding that:

Benchmark 4.3: Natural processes shape Earth's surface *(for example: landslides, weathering, erosion, mountain building, volcanic activity).*

Assessment Objective

a: Explain why Earth's surface is always building up in some places and wearing and down in others *(types of erosion, types of deposition).*

Student Strategies:

Question 69 is a two-point constructed-response question. The correct answer should include:

- two reasonable forces that change Earth's surface constantly.

69 **Earth is always changing.**

What are **two** forces that cause Earth's surface to constantly change?

_____

_____

_____

_____

_____

Go On ➡

Analysis: *Constructed-response answers may vary. The nature of Earth is to build from the inside and break down from the outside. The forces could include: plate tectonics, convection within the mantle of Earth, or weathering and erosion.*

Question **70** assesses:

Strand: **Science**

Standard 4: Earth and Space Science: Students know and understand the processes and interactions of Earth's systems and the structure and dynamics of Earth and other objects in space. *(Focus: Geology, Meteorology, Astronomy, Oceanography)*
Students know and can demonstrate understanding that:

Benchmark 4.4: Major geological events such as earthquakes, volcanic eruptions, and mountain building are associated with plate boundaries and attributed to plate motion.

Assessment Objective

a: Understand plate boundaries, their movements, and the resulting geologic events.

Student Strategies:

Tectonic plates are vast, moveable pieces of Earth's crust. These plates can shift and change due to volcanic or seismic activities. There are three basic types of seismic changes. The first is **extensional change**. Here, the plates shift away from one another. In the second type of seismic activity, **compressional change**, the plates shift toward one another. The last type of seismic activity occurs when plates transform, or shift alongside one another. Each type of change has different results on landforms above land below the surface of the Earth. **Transformations** cause less sinking and lifting, and the earthquakes generally have a magnitude of less than 8.5 on the Richter scale. Earthquakes along extensional fault lines tend to be shallow with a magnitude less than 8 on the Richter scale. Earthquakes along compressional boundaries can exceed a magnitude of 9 on the Richter scale. Mountain building, the creation of new mountains and mountain ranges, can also be caused by high-magnitude tectonic plate shifts or by volcanic eruptions.

70 The highest frequency of volcanic eruptions occurs around the "Ring of Fire" in the Pacific Ocean.

Why is this area so geologically active?

○ because there is a lot of energy in the size of the Pacific Ocean

○ because of spreading zones on the edge of the Pacific Ocean

○ because this is an area where several tectonic plates are crashing into each other

○ because the moon pulls harder on this portion of Earth

Go On

Analysis: *The third choice is correct. Volcanoes happen most frequently near the edges of tectonic plates that are colliding. The first choice is incorrect. This energy does not make volcanoes. The second choice is incorrect. The spreading zones are in the middle of the ocean bottom. The fourth choice is incorrect. The pull of the moon creates tides.*

Science Practice Tutorial — CSAP Science for Grade 8

Question **71** assesses:

Strand: **Science**

Standard 4: Earth and Space Science: Students know and understand the processes and interactions of Earth's systems and the structure and dynamics of Earth and other objects in space. *(Focus: Geology, Meteorology, Astronomy, Oceanography)*
Students know and can demonstrate understanding that:

Benchmark 4.5: Fossils are formed and used as evidence to indicate that life has changed through time.

Assessment Objective

a: Describe methods of fossil formation.

Student Strategies:

Question 71 is a two-point constructed-response question. The correct answer should include:

- two reasonable conditions necessary for fossil formation.

71 Describe **two** different conditions which must be present for fossils to form.

_____

_____

_____

_____

_____

Go On ▶

Analysis: *Constructed-response answers may vary. Most fossils are formed when a plant or an animal dies and is buried in mud and silt. Sediment builds over the mud and hardens into rock where it is not disturbed for thousands of years.*

Science Practice Tutorial   CSAP Science for Grade 8

Question **72** assesses:

Strand: **Science**

Standard 4: Earth and Space Science: Students know and understand the processes and interactions of Earth's systems and the structure and dynamics of Earth and other objects in space. *(Focus: Geology, Meteorology, Astronomy, Oceanography)*
Students know and can demonstrate understanding that:

Benchmark 4.6: Successive layers of sedimentary rock and the fossils contained within them can be used to confirm age, geologic time, history, and changing life forms of the Earth; this evidence is affected by the folding, breaking and uplifting of layers.

Assessment Objective

a: Interpret rock layers, including position *(concept of superpositioning)*, composition and fossil content to determine past conditions.

Student Strategies:

Fossils can be dated several ways. Fossils, like rocks, are generally deposited in the soil in horizontal layers, so fossils nearer the surface are usually younger than those found deeper down. Also, fossils are the same age as the rocks they are found in or around, so if the ages of the rocks around them can be determined, so can the age of the fossil.

CSAP Science for Grade 8   Science Practice Tutorial

72 Examine the diagram of rock layers below.

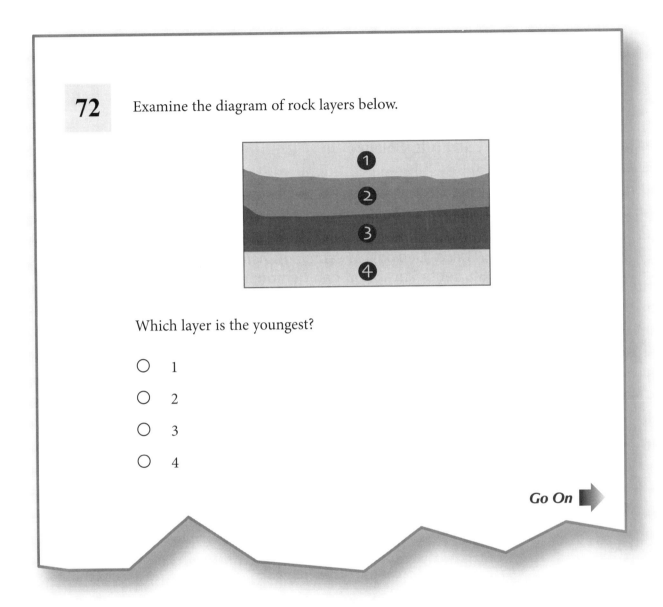

Which layer is the youngest?

○ 1

○ 2

○ 3

○ 4

Go On ▶

Analysis: *The first choice is correct. According to the concept of superpositioning, younger layers are generally formed on top of older layers. Layer 1 is the upper-most layer in this drawing. The second choice is incorrect. Layer 2 is below layer 1. The third choice is incorrect. Layer 3 is below layer 1. The fourth choice is incorrect. Layer 4 is below layer 1.*

Science Practice Tutorial   CSAP Science for Grade 8

Question **73** assesses:

Strand: **Science**

Standard 4: Earth and Space Science: Students know and understand the processes and interactions of Earth's systems and the structure and dynamics of Earth and other objects in space. *(Focus: Geology, Meteorology, Astronomy, Oceanography)*
Students know and can demonstrate understanding that:

Benchmark 4.6: Successive layers of sedimentary rock and the fossils contained within them can be used to confirm age, geologic time, history, and changing life forms of the Earth; this evidence is affected by the folding, breaking and uplifting of layers.

Assessment Objective

b: Predict the change in rock layer sequence due to folding, breaking and uplifting.

Student Strategies:

Question 73 is a two-point constructed-response question. The correct answer should include the events ordered from oldest to most recent.

- 2 points—all correct
- 1 point—two or three correct
- 0 points—one correct or none correct

73 Examine the diagram of rock layers below.

- cobbles and boulders
- volcanic ash
- erosion of the mountain
- mountain

Order the events below from oldest (1) to most recent (4).

_____ scattering of the boulders by a river

_____ uplifting and folding of the mountain

_____ erosion of the mountain

_____ ash flow from the volcano

Go On

Analysis: *Sometimes superposition gets complicated. Just read the layers from oldest to youngest—bottom to top. The layers are numbered 4, 1, 2, 3 from top to bottom. The oldest event is the folding and uplifting of the mountain, followed by the erosion of the mountain, and ash flow from the volcano. The youngest event is the scattering of boulders by a river.*

Question **74** assesses:

Strand: **Science**

Standard 4: Earth and Space Science: Students know and understand the processes and interactions of Earth's systems and the structure and dynamics of Earth and other objects in space. *(Focus: Geology, Meteorology, Astronomy, Oceanography)*
Students know and can demonstrate understanding that:

Benchmark 4.7: The atmosphere has basic composition, properties, and structure *(for example: the range and distribution of temperature and pressure in the troposphere and stratosphere).*

Assessment Objective

a: Identify all of the layers of the atmosphere, their order and the properties and individual characteristics that define them.

Student Strategies:

The atmosphere is divided into layers. It is thickest near the surface and thins out with height until it eventually merges with space.

Troposphere: This layer starts at Earth's surface and extends up to 10 miles high. The troposphere is the densest layer of the atmosphere. The further away from Earth's surface you go in this layer, the colder it becomes. Almost all weather occurs in this layer.

Stratosphere: This layer starts just above the troposphere and extends to 30 miles high. The stratosphere is drier than the troposphere and less dense. The temperature in this layer increases the higher you go because of the absorption of radiation from the sun. The ozone layer is in this layer.

Mesosphere: This layer starts just above the stratosphere and extends 50 miles high. In the mesosphere, temperatures decrease the higher you go.

Thermosphere: This layer starts just above the mesosphere and extends up to almost 375 miles high. The temperatures in the thermosphere increase as you climb in altitude because of the sun's energy.

74 Study the diagram of Earth's atmosphere below.

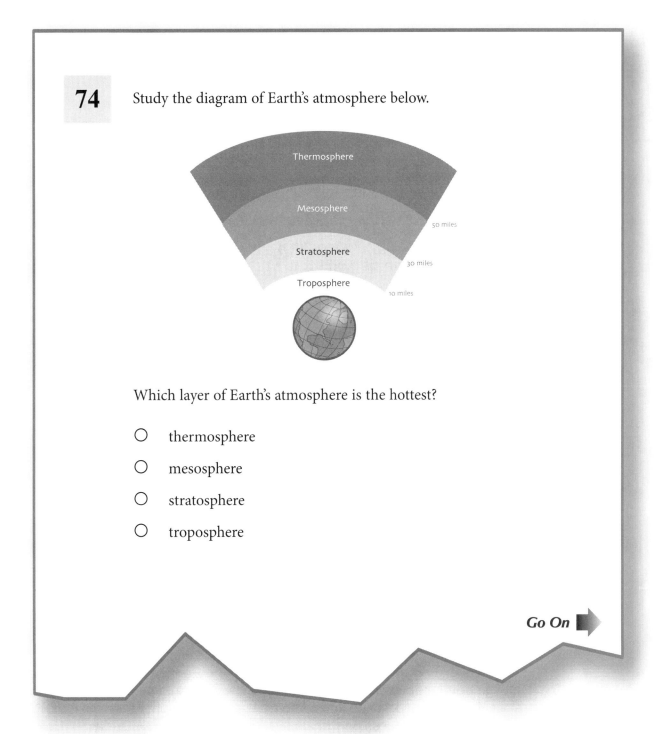

Which layer of Earth's atmosphere is the hottest?

○ thermosphere

○ mesosphere

○ stratosphere

○ troposphere

Go On

Analysis: *The first choice is correct. The thermosphere is the upper-most layer and closest to space (the sun). Generally, it gets colder as you get higher in the troposphere and the mesosphere. The opposite is true in the other layers. The second choice is incorrect. The mesosphere is very cold. The third choice is incorrect. Although it warms, the stratosphere does not get much above freezing. The fourth choice is incorrect. The troposphere is the layer we live in.*

Question **75** assesses:

Strand: **Science**

Standard 4: Earth and Space Science: Students know and understand the processes and interactions of Earth's systems and the structure and dynamics of Earth and other objects in space. *(Focus: Geology, Meteorology, Astronomy, Oceanography)* **Students know and can demonstrate understanding that:**

Benchmark 4.8: Atmospheric circulation is driven by solar heating *(for example: the transfer of energy by radiation, convection, conduction).*

Assessment Objective

a: Explain that the Sun heats Earth via radiation that in turn heats the atmosphere via conduction and convection.

Student Strategies:

Energy from the sun reaches the Earth through radiation. Due to the absorption, reflection, and scattering of different wavelengths of radiation, less than half of the solar radiation that enters the atmosphere reaches Earth's surface and only about one-fifth warms the atmosphere directly.

Most of the energy that warms Earth's atmosphere comes indirectly from the heated Earth. Some of the energy absorbed by the Earth warms the atmosphere through **conduction**, which is the transfer of heat within a substance. However, since the gases and liquids that comprise Earth's atmosphere are not good conductors, this process only heats the air just above Earth's surface. Most of the heat in the atmosphere comes from **convection**, which is the vertical transfer of energy by the movement of the heated substance. The warm air from the surface of the Earth rises, displacing the cooler air above, which sinks toward the Earth and is heated and rises. Conduction and convection are called **sensible heat transfer processes**.

75 The heat from the sun comes to Earth in which of the following forms?

○ conduction

○ convection

○ friction

○ radiation

Go On ➤

Analysis: *The fourth choice is correct. The energy from the sun can only travel though space as radiation. The atmosphere heats up as a result of convection in the air. The first choice is incorrect. Conduction requires contact with the object heating it. The second choice is incorrect. Convection requires a substance to circulate. There isn't anything to circulate in space. The third choice is incorrect. Friction comes from mechanical contact.*

Question **76** assesses:

Strand: **Science**

Standard 4: Earth and Space Science: Students know and understand the processes and interactions of Earth's systems and the structure and dynamics of Earth and other objects in space. *(Focus: Geology, Meteorology, Astronomy, Oceanography)*
Students know and can demonstrate understanding that:

Benchmark 4.9: There are quantitative changes in weather conditions over time and space *(for example: humidity, temperature, air pressure, cloud cover, wind, precipitation)*.

Assessment Objective

a: Interpret weather data and the changes that occur over time *(graph, charts, weather maps)*.

Student Strategies:

Question 76 is a two-point constructed-response question. The correct answer should include:

- an explanation of the circulation of air around high pressure systems
- an explanation of the circulation of air around low pressure systems.

76 Examine the weather map below.

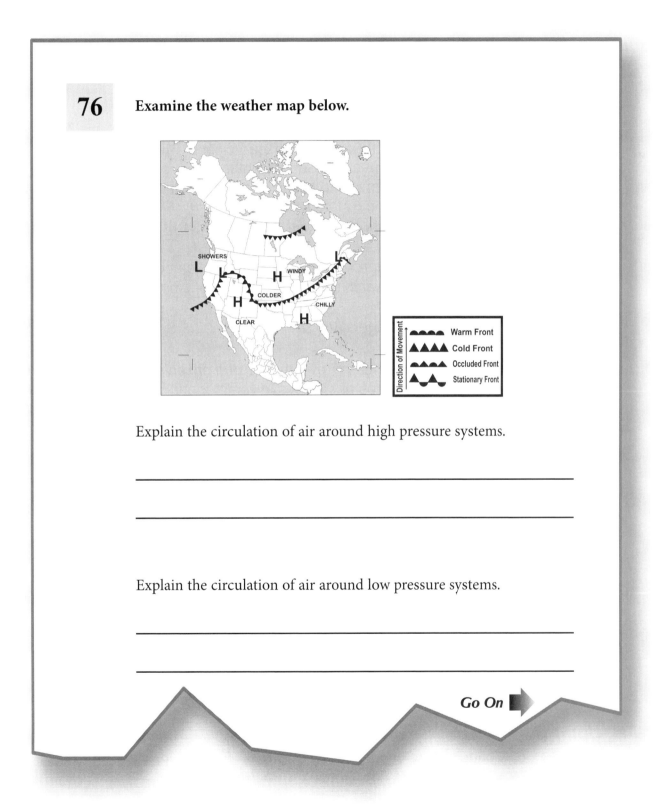

Explain the circulation of air around high pressure systems.

_____

_____

Explain the circulation of air around low pressure systems.

_____

_____

Go On

Analysis: *Constructed-response answers may vary. Air circulates clockwise around a high pressure system. Air circulates counterclockwise around a low pressure system.*

Question **77** assesses:

Strand: **Science**

Standard 4: Earth and Space Science: Students know and understand the processes and interactions of Earth's systems and the structure and dynamics of Earth and other objects in space. *(Focus: Geology, Meteorology, Astronomy, Oceanography)*
Students know and can demonstrate understanding that:

Benchmark 4.10: There are large-scale and local weather systems *(for example: fronts, air masses, storms)*.

Assessment Objective

a: Use several pieces of evidence *(cloud observations, weather maps)* to identify causes of changes in weather and weather patterns *(weather moves west to east)*.

Student Strategies:

Clouds can be characterized as **cirrus** (high clouds), **alto** (middle clouds), **stratus** (low clouds), **cumulus** (clouds with vertical growth), and **special clouds**.

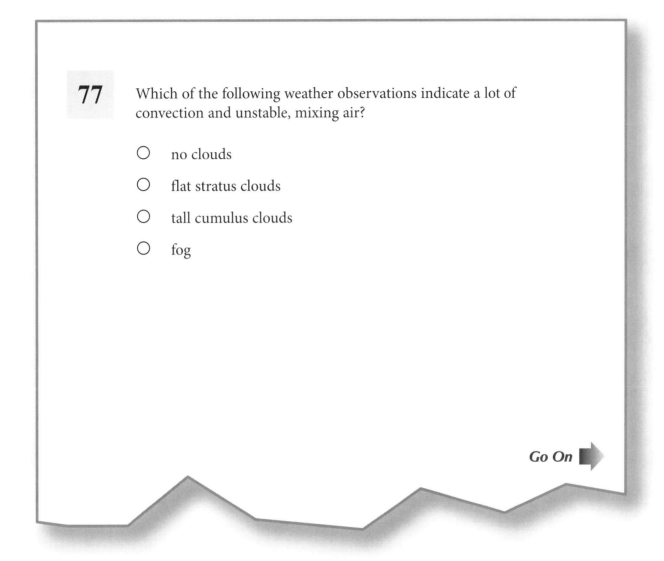

77 Which of the following weather observations indicate a lot of convection and unstable, mixing air?

○ no clouds

○ flat stratus clouds

○ tall cumulus clouds

○ fog

Go On ▶

Analysis: *The third choice is correct. Tall cumulus clouds indicate a lot of convection and very unstable air and could bring a storm. The first choice is incorrect. It is hard to observe the air without clouds. The second choice is incorrect. The air is stable when flat stratus clouds are observed. The fourth choice is incorrect. Fog indicates stable air.*

Question **78** assesses:

Strand: **Science**

Standard 4: Earth and Space Science: Students know and understand the processes and interactions of Earth's systems and the structure and dynamics of Earth and other objects in space. *(Focus: Geology, Meteorology, Astronomy, Oceanography)*
Students know and can demonstrate understanding that:

Benchmark 4.10: There are large-scale and local weather systems *(for example: fronts, air masses, storms).*

Assessment Objective

b: Identify the inter-relationship between large scale weather systems and local weather.

Student Strategies:

Question 78 is a two-point constructed-response question. The correct answer should include:

- two ways that jet stream winds affect weather.

78 Explain **two** ways that jet stream winds affect our local weather.

_____

_____

_____

_____

_____

Go On ➡

Analysis: *Constructed-response answers may vary. Jet stream winds are strong winds that are at an altitude higher than 20,000 feet above Earth's surface. Jet streams follow the boundaries between hot and cold air. Since these hot and cold boundaries are more pronounced in winter, jet stream winds are stronger during the winter and weaker in the summer. A correct answer would include that jet stream drives our weather from west to east and can pull cold air down from the Arctic or drive warm air north from southern North America. It can also drive high altitude moisture around us.*

Question **79** assesses:

Strand: **Science**

Standard 4: Earth and Space Science: Students know and understand the processes and interactions of Earth's systems and the structure and dynamics of Earth and other objects in space. *(Focus: Geology, Meteorology, Astronomy, Oceanography)*
Students know and can demonstrate understanding that:

Benchmark 4.10: There are large-scale and local weather systems *(for example: fronts, air masses, storms)*.

Assessment Objective

c. Explain how Earth's surface features *(such as mountains, oceans)* affect local weather.

Student Strategies:

Question 79 is a two-point constructed-response question. The correct answer should include:

- the cause of upslope winds;
- when upslope winds occur.

79 One of the local weather effects we experience almost daily, especially in the mountains, is called upslope winds.

What causes these winds and when do they occur?

_____

_____

_____

_____

Go On ➡

Analysis: *Constructed-response answers may vary. Upslope winds are the effect of convection and are stronger when local geography has hillsides and mountains to channel the wind. The opposite wind happens at night—downslope winds. The cause is convection from heating of the soil at the foot of a hill or mountain. These winds happen during the day and get stronger in the late afternoon when the sun has been heating the soil for a long time.*

Question **80** assesses:

Strand: **Science**

Standard 4: Earth and Space Science: Students know and understand the processes and interactions of Earth's systems and the structure and dynamics of Earth and other objects in space. *(Focus: Geology, Meteorology, Astronomy, Oceanography)*
Students know and can demonstrate understanding that:

Benchmark 4.11: The world's water is distributed and circulated through oceans, glaciers, rivers, groundwater, and atmosphere.

Assessment Objective

a: Explain the processes and relationships that connect elements *(all water sources)* of the water cycle.

Student Strategies:

In the water cycle, the sun is largely responsible for the interaction between matter and energy. The sun heats up bodies of water, including lakes, rivers, and oceans. As the water warms, it becomes water vapor and evaporates into the air. At the same time, plants are losing water through their leaves in a process called transpiration. This water also becomes water vapor and evaporates into the air. As water vapor collects in the air, it condenses and forms clouds. Here, the water vapor is cooled and once again changes into liquid form. As the clouds become larger, they also become heavier, and can no longer hold the condensed water. At this point, precipitation occurs. Water falls to Earth in the form of snow, sleet, hail, or rain. Once the water falls to Earth, it is collected in the groundwater system. It once again becomes part of the water collection at the surface of Earth, such as rivers, lakes, and oceans. The sun will once again heat the water until some of it evaporates, beginning the water cycle again. Without energy from the sun, the water cycle would not occur.

80 Examine the drawing of the water cycle below.

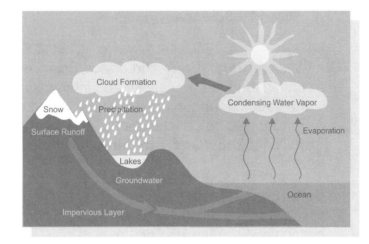

What form of energy drives water through the water cycle?

- ○ gravity
- ○ solar energy
- ○ geothermal energy
- ○ potential energy

Go On

Analysis: *The second choice is correct. The sun provides the energy to evaporate water and to move the water around by convective heating of the atmosphere. The first choice is incorrect. Gravity only helps rain fall. The third choice is incorrect. Geothermal energy is deep in the earth. The fourth choice is incorrect. At some point, the water gains potential energy, but the water cycle is driven by the effects of solar energy.*

Question **81** assesses:

Strand: **Science**

Standard 4: Earth and Space Science: Students know and understand the processes and interactions of Earth's systems and the structure and dynamics of Earth and other objects in space. *(Focus: Geology, Meteorology, Astronomy, Oceanography)*
Students know and can demonstrate understanding that:

Benchmark 4.12: The ocean has a certain composition and physical characteristics *(for example: currents, waves, features of the ocean floor, salinity, and tides).*

Assessment Objective

a: Understand the composition and physical characteristics of oceans *(for example: temperature, salinity, wavelength, ocean floor, etc).*

Student Strategies:

Approximately 71% of Earth's surface is covered by oceans. More than half of this area is over 3,000 meters (9,800 ft) deep. Average oceanic salinity is around 3.5%, and nearly all seawater has salinity in the range of 3.1% to 3.8%.

The average depth of the ocean is 3,790 meters (12,430 ft), but this varies greatly depending on the ocean floor. Some areas, such as the continental shelf, are much more shallow than other areas, such as a deep ocean trench.

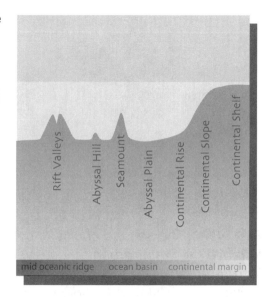

81 How does the composition of sea water compare to the composition of human cells?

○ They are identical.

○ They have the same components in about the same amounts.

○ They are very different and have very different components.

○ They might have been the same once but they are not even close now.

Go On

Analysis: *The second choice is correct. The composition of human cells and sea water are very close to one another. Human cells consist of 65–90% water (H_2O) and oxygen makes up 65% of cell mass. Sea water contains 85.84% oxygen and approximately 96% of H_2O. Human cells and sea water also contain carbon, hydrogen, chlorine, calcium, potassium, and other similar components. The first choice is incorrect. There are slight insignificant differences between human cells and sea water. The third choice is incorrect. The composition of human cells and sea water are very close to one another. The fourth choice is incorrect. The composition of human cells and sea water are very close to one another.*

Question **82** assesses:

Strand: **Science**

Standard 4: Earth and Space Science: Students know and understand the processes and interactions of Earth's systems and the structure and dynamics of Earth and other objects in space. *(Focus: Geology, Meteorology, Astronomy, Oceanography)*
Students know and can demonstrate understanding that:

Benchmark 4.13: There are characteristics *(components, composition, size)* and scientific theories of origin of the Solar System.

Assessment Objective

a: Describe the parts *(planets, Sun, moons, asteroids, comets)* of the Solar System and their motions.

Student Strategies:

The solar system is made up of all the objects that orbit our sun. The sun is easily the largest object in the solar system. Below is an image that shows the relative sizes of the planets and the sun.

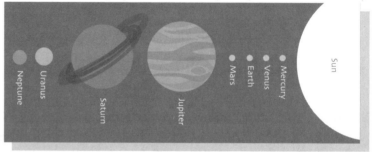

The relative sizes of all eight planets in our solar system (distances are not to scale)

The moons of each planet are much smaller than the planets themselves. Almost all moons are thought to come from the same material as the planets they orbit. Remember, most of the inner planets in our solar system are small and rocky, while most of the outer planets are very large and made of gases.

82 What is the shape of the orbits of the planets in our solar system?

○ circular

○ triangular

○ elliptical

○ Each planet's orbit is a different shape.

Go On

Analysis: *The third choice is correct. The shape of the orbits of the planets is determined by the sun's gravity and the planets' distance from the sun. The orbits of the planets in our solar system are elliptical. The first choice is incorrect. The orbits of the planets in our solar system are elliptical, not circular. The second choice is incorrect. The orbits of the planets in our solar system are elliptical, not triangular. The fourth choice is incorrect. The orbits of the planets in our solar system are elliptical.*

Science Practice Tutorial　　　　　　　　　　　　　　　CSAP Science for Grade 8

Question **83** assesses:

Strand: **Science**

Standard 4: Earth and Space Science: Students know and understand the processes and interactions of Earth's systems and the structure and dynamics of Earth and other objects in space. *(Focus: Geology, Meteorology, Astronomy, Oceanography)*
Students know and can demonstrate understanding that:

Benchmark 4.13: There are characteristics *(components, composition, size)* and scientific theories of origin of the Solar System.

Assessment Objective

b: Compare and contrast the characteristics of the Sun, Moon and Earth.

Student Strategies:

Question 83 is a two-point constructed-response question. The correct answer should include:

- a sketch of the sun, Earth, and Earth's moon
- a description of the motions of the sun, Earth, and Earth's moon

83 Describe or sketch the motions of the sun, Earth, and Earth's moon.

Analysis: *Constructed-response answers may vary. The description or sketch should show that Earth orbits the sun and the moon orbits Earth.*

Question **84** assesses:

Strand: **Science**

Standard 4: Earth and Space Science: Students know and understand the processes and interactions of Earth's systems and the structure and dynamics of Earth and other objects in space. *(Focus: Geology, Meteorology, Astronomy, Oceanography)*
Students know and can demonstrate understanding that:

Benchmark 4.13: There are characteristics *(components, composition, size)* and scientific theories of origin of the Solar System.

Assessment Objective

c: Examine and explain the scientific theories on the formation of our Solar System, Earth, and Moon.

Student Strategies:

Question 84 is a two-point constructed-response question. The correct answer should include:

- a detailed theory about the formation of the moon;
- why the theory is reasonable.

84 Following the Big Bang and the formation of the solar system, our moon also formed though no one really knows how.

Descibe **one** theory for the formation of the moon.

_____

_____

_____

_____

_____

_____

_____

_____

_____

Go On ➡

Analysis: *Constructed-response answers may vary. There are three current theories about the formation of the moon. The answer should describe a reasonable theory such as: a part of Earth spun out into space, the moon formed from a cloud of matter like Earth, or Earth captured a passing body in its gravitational field.*

Question **85** assesses:

Strand: **Science**

Standard 4: Earth and Space Science: Students know and understand the processes and interactions of Earth's systems and the structure and dynamics of Earth and other objects in space. *(Focus: Geology, Meteorology, Astronomy, Oceanography)*
Students know and can demonstrate understanding that:

Benchmark 4.14: Relative motion, axes tilt and positions of the Sun, Earth, and Moon have observable effects *(for example: seasons, eclipses, moon phases).*

Assessment Objective

a: Understand how the location of the Moon affects the phases of the Moon, eclipses, and the tides.

Student Strategies:

It takes the moon approximately 28 days to revolve around Earth. Although the view of the moon is constantly changing, there are eight noticeable phases of the moon. Beginning with a **full moon**, where the moon is fully reflecting the sun's light, the moon will wane into the other phases. The next phases are the **waning gibbous**, the **third quarter**, and the **waning crescent**. The next phase, the **new moon**, is the point at which the moon cannot be viewed from the night sky because no sunlight is reflected from its surface. After this point, the moon begins to wax (appear larger), starting with the **waxing crescent**. The **first quarter** and the **waxing gibbous** are the final two phases before the moon returns back to the full moon phase. Since the moon spins facing Earth in the same direction, we only see about half of the surface of the moon at any time, even during a full moon. The other half of the moon is always facing away from Earth.

Sometimes Earth and the moon block one another from the light of the sun. When the moon blocks Earth from receiving the sun's light, it is called a **solar eclipse**. The moon is much smaller than Earth, so the solar eclipse will only put a small portion of Earth in a shadow. One small area will be completely shadowed by a total solar eclipse, another area will experience a partial solar eclipse, and most of the Earth will not be affected at all. At other times, Earth blocks the moon from receiving light from the sun. When the moon is hidden from the light of the sun by the shadow of the Earth, it is called a **lunar eclipse**. Since Earth is much larger than the moon, the entire moon is in the shadow of Earth during a lunar eclipse.

85 On a monthly cycle, the moon goes through different phases in which we see it as lit up differently from new moon to full moon.

What are the phases of the moon are caused by?

○ the shadow of Earth

○ the angle of our view as the moon orbits Earth

○ monthly eclipses

○ differential lighting by the sun on the moon

Go On ▶

Analysis: *The second choice is correct. The angle of our view as the moon orbits Earth causes the phases of the moon. The first choice is incorrect. The shadow of Earth causes eclipses, not phases. The third choice is incorrect. Eclipses are not monthly. The fourth choice is incorrect. The sun shines the same on the moon all of the time.*

Science Practice Tutorial — CSAP Science for Grade 8

Question **86** assesses:

Strand: **Science**

Standard 4: Earth and Space Science: Students know and understand the processes and interactions of Earth's systems and the structure and dynamics of Earth and other objects in space. *(Focus: Geology, Meteorology, Astronomy, Oceanography)*
Students know and can demonstrate understanding that:

Benchmark 4.14: Relative motion, axes tilt and positions of the Sun, Earth, and Moon have observable effects *(for example: seasons, eclipses, moon phases)*.

Assessment Objective

b: Understand how the tilt and motions of Earth results in days, years and seasons.

Student Strategies:

Our solar system is comprised of the sun, eight planets and their moons, and the asteroid belt. All of the eight planets revolve around the sun in different orbital paths from one another. An Earth year, 365 1/4 days, is the amount of time it takes Earth to revolve completely around the sun.

In addition to revolving around the sun, Earth also rotates on its axis. Earth's axis is not straight up and down; Earth tilts at about a 23° angle from a perpendicular position. Daylight and darkness are caused by the spinning of Earth on its axis. It takes 24 hours for Earth to spin completely around one time.

The seasons are also caused by the tilt of Earth. The geographic regions along the equator do not have "true seasons" with a great variance in temperature and daylight. The reason for this is that geographic regions of Earth at the equator remain at a fairly constant angle from the sun. The areas along the equator receive fairly equal amounts of light and heat energy from the sun throughout the year.

Other regions, outside of tropical zones, have a greater variation of light from the sun, and therefore a greater variance in seasonal weather patterns. The seasons also vary according to the hemisphere. For instance, when the Northern Hemisphere is tilted toward the sun, it is summer there. At the same time, the Southern Hemisphere will be tilted away from the sun, and it will be winter there.

86 **The position and tilt of Earth give it particular motions and effects.**

Which of the following is the effect of the tilt of Earth toward or away from the sun?

○ the passage of day and night

○ the passage of the seasons

○ the phases of the moon

○ global warming

Go On ▶

Analysis: *The second choice is correct. As Earth tilts toward the sun in May, June, and July, it gets more direct sunlight and is therefore warmer in the Northern Hemisphere. The opposite is true in November, December, and January. The first choice is incorrect. The spin of Earth on its axis creates day and night. The third choice is incorrect. The phases of the moon are caused by the orbit of the moon around Earth. The fourth choice is incorrect. Global warming is not the result of any of the motions of Earth.*

Question **87** assesses:

Strand: **Science**

Standard 4: Earth and Space Science: Students know and understand the processes and interactions of Earth's systems and the structure and dynamics of Earth and other objects in space. *(Focus: Geology, Meteorology, Astronomy, Oceanography)*
Students know and can demonstrate understanding that:

Benchmark 4.15: The universe consists of many billions of galaxies *(each containing many billions of stars)* and that vast distances separate these galaxies and stars from one another and from Earth.

Assessment Objective

a: Describe the components of the universe in terms of galaxies, stars, and solar systems.

Student Strategies:

All stars in our universe are made of hydrogen that is fusing to become helium. However, stars can be different sizes, colors, temperatures, and ages. They often have different characteristics because of these differences.

The color of a star varies from red to blue as the star becomes hotter. Our sun is a very average star in most respects, including size. The smallest stars may be as small as 1/10 the size of our sun and the largest may be a thousand times as massive.

As a star ages, it burns through its hydrogen fuel and eventually collapses into a smaller form called a **white dwarf**. If the star is more massive, it may explode into a **supernova** before collapsing into a **neutron star**. However, if a star is very massive, it may instead form a **black hole** after collapsing.

Our sun is the center of our solar system. Our solar system is one of billions, or even trillions, of stars in the Milky Way galaxy. A **galaxy** is a collection of stars that move as a group. The universe is filled with millions of galaxies. These galaxies are held together by gravity.

87 As far as we know, the universe is composed of many different kinds of large and distant objects, including galaxies, stars, and nebulae.

Which statement is **true** about these objects in the universe?

○ They are all about the same distance from Earth.

○ They are all about the same size.

○ They are all moving away from about the same place in the universe.

○ They are all moving quickly toward the same spot in the universe.

Go On ▶

Analysis: *The third choice is correct. Edwin Hubble actually calculated that these objects are all moving away from a central point or place in the universe. It is the result of the Big Bang. The first choice is incorrect. The distances of these objects from Earth vary greatly. The second choice is incorrect. The sizes of these objects also vary greatly. The fourth choice is incorrect. This is the opposite of our observations.*

Science Practice Tutorial   CSAP Science for Grade 8

Question **88** assesses:

Strand: **Science**

Standard 4: Earth and Space Science: Students know and understand the processes and interactions of Earth's systems and the structure and dynamics of Earth and other objects in space. *(Focus: Geology, Meteorology, Astronomy, Oceanography)*
Students know and can demonstrate understanding that:

Benchmark 4.16: Technology is needed to explore space *(for example: telescopes, spectroscopes, spacecraft, life support systems)*.

Assessment Objective

a: Understand the technologies needed to explore space and evaluate their effectiveness and challenges.

Student Strategies:

Question 88 is a two-point constructed-response question. The correct answer should include:

- two technologies that came from space exploration that we use in everyday life;
- what the technologies are used for.

88 Space exploration has resulted in many technologies we use in everyday life.

Name **two** of these technologies we use.

_____

_____

_____

_____

_____

Go On

Analysis: *Constructed-response answers may vary. There are many new materials that have resulted from space exploration. Correct answers will list two reasonable technologies that probably came from the space program, like Velcro, graphite coatings, enriched baby foods, athletic shoes, smoke detectors, freeze-dried technology, and more.*

Question **89** assesses:

Strand: **Science**

Standard 5: Students understand that the nature of science involves a particular way of building knowledge and making meaning of the natural world. **Students know and can demonstrate understanding that:**

Benchmark 5.1: A controlled experiment must have comparable results when repeated.

Assessment Objective

a: Identify a controlled factor in a scientific investigation.

Student Strategies:

Question 89 is a two-point constructed-response question. The correct answer should include:

- reason for controls in an experiment;
- why controls are important in an experiment.

89 **A control in an experiment is defined as a variable kept the same in all parts of the experiment.**

Why is a control important in an experiment?

_____

_____

_____

_____

_____

_____

_____

Go On ➡

Analysis: *Constructed-response answers may vary. The idea in a controlled experiment is to test only one variable at a time. Correct answers will describe holding the controlled variables all the same in an experiment and testing only one variable; i.e. having only one thing change in the trials.*

Question **90** assesses:

Strand: **Science**

Standard 5: Students understand that the nature of science involves a particular way of building knowledge and making meaning of the natural world. **Students know and can demonstrate understanding that:**

Benchmark 5.1: A controlled experiment must have comparable results when repeated.

Assessment Objective

b: Explain that by repeating a controlled experiment, it should lead to comparable results.

90 What is the effect of repeating an experiment?

○ to see if you made any errors in your first experiment

○ to see how varied the results of the experiment can be

○ to develop comparable results

○ All of the above are correct.

Go On

Analysis: *The fourth choice is correct. One of the qualifications of scientific knowledge is that it should be repeatable. Checking for errors, validity of the experiment, and developing comparable results are all good reasons to repeat an experiment.*

Question **91** assesses:

Strand: **Science**

Standard 5: Students understand that the nature of science involves a particular way of building knowledge and making meaning of the natural world. **Students know and can demonstrate understanding that:**

Benchmark 5.1: A controlled experiment must have comparable results when repeated.

Assessment Objective

c: Identify and/or explain that evidence collected through repeated experiments cannot be accurately compared to previous experimental results, if conditions were not kept the same.

91 **Comparability between experiments is important if you want to develop confidence in your results.**

All of the following should remain the same when repeating an experiment **except**

○ the control variables.

○ the experimental method.

○ the results.

○ the experimental variable.

Go On

Analysis: *The third choice is correct. The results could come out a little different in a repeated experiment. The first, second, and fourth choices are incorrect. Everything should be the same in an experiment that is being repeated so you can compare the results.*

Question **92** assesses:

Strand: **Science**

Standard 5: Students understand that the nature of science involves a particular way of building knowledge and making meaning of the natural world. **Students know and can demonstrate understanding that:**

Benchmark 5.2: Scientific knowledge changes as new knowledge is acquired and previous ideas are modified *(for example: through space exploration)*.

Assessment Objective

a: Identify and/or describe the reasons why scientific knowledge changes over time.

Student Strategies:

Science is not just a list of facts and laws. Science is always in motion. One of the principles of scientific study is that knowledge is available to all people and any scientific study is open to criticism or testing by other scientists.

No scientific knowledge provides the whole "truth." Instead, knowledge is tested and added to what scientists have learned in the past. This is why the efforts of scientists that made observations hundreds and even thousands of years ago are still very important. When a scientist makes a discovery, it is usually termed that he or she is "standing on the shoulders of giants." This means that the scientist could not have made this discovery without using the knowledge gained by many scientists in the past.

Past scientific knowledge is also important because it allows new scientists to look at older studies and knowledge with fresh perspective. Scientists who study older scientific theories and then offer new ideas make many scientific discoveries.

92 Which of the following could change scientific ideas over time?

○ more experiments

○ more observations over time

○ new technologies

○ All of the above are correct.

Go On

Analysis: *The fourth choice is correct. Science tends to develop over time because of more experiments, more observations, and new technologies.*

Question **93** assesses:

Strand: **Science**

Standard 5: Students understand that the nature of science involves a particular way of building knowledge and making meaning of the natural world. **Students know and can demonstrate understanding that:**

Benchmark 5.3: Contributions to the advancement of science have been made by people in different cultures and at different times in history.

Assessment Objective

a: Recognize the concept of multicultural contributions to the advancement of science over time.

93 Which of the following is **incorrect** about science?

○ Science is an uniquely American endeavor.

○ People from many cultures contribute to scientific ideas.

○ Biology is practiced by scientists from around the world.

○ Science has been practiced for thousands of years by people from many parts of the globe.

Go On

Analysis: *The first choice is correct. Science is not an uniquely American endeavor; science is a way of thinking that is not unique to a particular culture. The second, third, and fourth choices are incorrect. All of these statements are true.*

Science Practice Tutorial   CSAP Science for Grade 8

Question **94** assesses:

Strand: **Science**

Standard 5: Students understand that the nature of science involves a particular way of building knowledge and making meaning of the natural world. **Students know and can demonstrate understanding that:**

Benchmark 5.4: Models can be used to predict change *(for example: computer simulation, video sequence, stream table).*

Assessment Objective

a: Recognize and/or describe that models can be used to obtain information about scientific processes and/or objects that may be difficult to study.

Student Strategies:

Question 94 is a two-point constructed-response question. The correct answer should include:

- a scientific model;
- which scientific idea the model explains.

94 Which is **not** a model used to understand something scientifically?

- ○ computer simulations
- ○ controlled experiments
- ○ stream tables
- ○ video sequences

Go On ➡

Analysis: *The second choice is correct. Experiments are not models. The first choice is incorrect. A computer model is a simplified simulation of something in nature we can use to predict how something works. Computer simulations are models. The third choice is incorrect. Stream tables are box-like tables that usually include sand to model the effects of a stream on land. The fourth choice is incorrect. Video sequences are digital images gathered or created to model the effect of variables on one another in an experiment.*

Question **95** assesses:

Strand: **Science**

Standard 5: Students understand that the nature of science involves a particular way of building knowledge and making meaning of the natural world. **Students know and can demonstrate understanding that:**

Benchmark 5.4: Models can be used to predict change *(for example: computer simulation, video sequence, stream table).*

Assessment Objective

b: Describe a model that would be appropriate to understand a scientific process and content.

Student Strategies:

Many times, when conducting experiments, scientists must use models of the natural world. Models are used for a variety of reasons. For instance, if scientists were testing a theory about ocean pollution, it might require the scientist to test the idea by creating a replica of the ocean. This "mini ocean" has certain limitations though, because it is not the real ocean. The scientist may be able to replicate the action of the waves, but it will be very difficult to recreate the depth of the ocean. This may or may not make a big difference in the outcome of the experiment, but it is important for students to know that scientific investigations are often limited when models are used. Models may be required and may be useful, but models are never a true replacement for the natural world. Students should be taught to consider how using a model would alter the conclusions that can be made from the investigation.

95 We use models to show scientific ideas all the time.

Describe **one** scientific model you know about and tell what scientific idea it explains.

_____

_____

_____

_____

_____

_____

Go On

Analysis: *Constructed-response answers may vary. A correct answer will name a model like the electronic cloud model of the atom and tell what it explains—how atoms bond together.*

Science Practice Tutorial — CSAP Science for Grade 8

Question **96** assesses:

Strand: **Science**

Standard 5: Students understand that the nature of science involves a particular way of building knowledge and making meaning of the natural world. **Students know and can demonstrate understanding that:**

Benchmark 5.4: Models can be used to predict change *(for example: computer simulation, video sequence, stream table).*

Assessment Objective

c: Explain that models are used to understand processes and predict change in many situations:

- where it may take several years to collect the data firsthand *(e.g., sea floor spreading, etc.)*
- where the event has already occurred and evidence has been lost or is limited *(e.g., asteroid impact, fossil record, etc.)*
- when a process is dangerous to study *(e.g., volcanoes, earthquakes, tornados, etc.)*
- when a process is very slow *(e.g., erosion, continental drift, rock cycle, climate change, etc.)*
- when the scale of size is difficult to replicate and makes observations difficult *(e.g., atoms, cells, solar system, etc.)*
- to make an abstract more understandable *(e.g., Newton's Laws and amusement park physics, etc.)*

Student Strategies:

Models can be helpful, but understanding the fundamental differences between models and natural world examples is important in drawing valid scientific conclusions. Sometimes the differences are obvious, but at other times understanding the difference between models and their natural world representations takes a great deal of careful thought.

96 When might a model have an advantage over direct observations?

○ when processes are very slow

○ when processes are too large to see

○ when processes are too small to see

○ All of the above are correct.

Go On ➡

Analysis: *The fourth choice is correct. There are many situations when a model might have advantages over observing directly from nature. All of these situations provide an advantage to using a model.*

Science Practice Tutorial CSAP Science for Grade 8

Question **97** assesses:

Strand: **Science**

Standard 5: Students understand that the nature of science involves a particular way of building knowledge and making meaning of the natural world. **Students know and can demonstrate understanding that:**

Benchmark 5.5: There are interrelationships among science, technology and human activity that affect the world.

Assessment Objective

a: Explain that human activity, including current scientific studies and technological advancements, can have both positive and negative effects on the natural world.

Student Strategies:

Question 97 is a two-point constructed-response question. The correct answer should include:

- a scientific activity;
- an advantage of that activity for the natural world.

97 Choose a scientific activity and tell **one** advantage of the activity for the natural world.

_____

_____

_____

_____

_____

Go On ▶

Analysis: *Constructed-response answers may vary. A correct answer will name a scientific activity, like the development of solar energy, and cite one reasonable advantage the activity has for the natural world, like reducing mining or producing more electricity.*

This page left intentionally blank.

Science Assessment One

Directions for Science Assessment One

This Grade 8 CSAP Science Assessment (Session One, Session Two, and Session Three) has multiple-choice and constructed-response questions. There are several important things to remember as you take the Science CSAP:

- Only No. 2 pencils may be used on any part of the test materials. This includes the front cover.
- During testing, no one is allowed to have electronic communication devices in the testing room.
- Read each multiple-choice question carefully. Think about what is being asked. Then fill in one answer bubble to mark your answer.
- If you do not know the answer to a multiple-choice question, skip it and go on.
- You **may** use the space provided in the test book. You **must** show all of your work in the space or on the lines provided.
- For constructed-response questions, write your response clearly and neatly on the lines provided.
- If you finish a session early, you may go back and check over your work on that session only. You may be allowed to read, but you are **not** allowed to write.

Science Assessment One—Session One

1 Which organ is part of the nervous system?

○ pancreas
○ cerebrum
○ trachea
○ duodenum

2 Which particle has a negative charge and is found outside the nucleus of an atom?

○ electron
○ neutron
○ prion
○ proton

3 If a scientist finds evidence that does not support a hypothesis, what happens to the hypothesis?

○ The hypothesis is thrown out.
○ The hypothesis is tested more.
○ A new hypothesis is automatically started.
○ The evidence is ignored because there already is a hypothesis.

Go On

4 Human body systems work together to accomplish different functions for the body. The circulatory system works with many other systems.

Name **one** body system that works with the circulatory system and describe what it does.

_____

_____

_____

_____

_____

_____

5 Earth is always changing.

What are **two** ways in which weathering changes the surface of Earth?

_____

_____

_____

_____

_____

Go On

6 Look at the thermometer below.

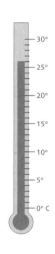

What is the temperature shown on the thermometer?

○ 26 degrees C

○ 2.6 degrees C

○ 26 degrees F

○ 2.6 degrees F

7 **One way of organizing how we understand the world is by looking at different levels of organization.**

What level within an organism is larger than a biome?

○ molecule

○ biosphere

○ organ

○ organism

8 Mountains are often formed where continental plates collide.

Which plates collided to create the Himalayas?

○ North and South America
○ Africa and Europe
○ Antarctica and Australia
○ India and Asia

9 There are at least two places on Earth's surface where continents seem to fit together like puzzle pieces. This was seen by Alfred Wegener in the early 20th century. His proposal was that a large supercontinent once existed.

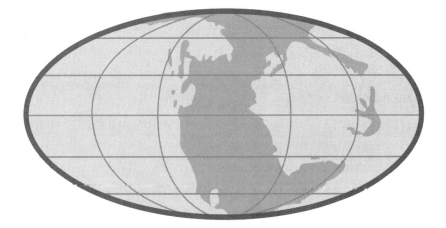

Explain **two** pieces of additional evidence that could help support his theory.

_____
_____
_____
_____

Go On

10 The theory of plate tectonics was not widely accepted in the early 20th century because evidence of what was moving the plates could not be found. Finally, beginning in the 1950s, new technology helped find evidence that supported this theory.

What new evidence was found?

○ Exploration of the ocean floor found sea floor spreading.

● Maps of the old continents were found.

○ A scientist saw it in a dream.

○ New computer models figured it out.

11 One physical characteristic that is used to classify vertebrates is whether or not an animal is a homeotherm. Homeotherms control their own body temperature and poikilotherms are the same temperature as their environment.

Which of the following are homeotherms?

○ fish and amphibians

○ amphibians and reptiles

○ reptiles and birds

● birds and mammals

12 Which property would be different enough to help separate a mixture of salt and white sand?

○ color

○ size

○ solubility

○ magnetic properties

13 A student was using a tool to identify a flower he found on a school trip. Here is a portion of the tool.

> 16. _____
>
> a. Flower is blue .... Blue Columbine
>
> b. Flower is not blue .... go to 17

Which of the following is this tool?

○ a field guide

○ a key

○ a manual

○ an ID card

14 Scientists are testing a new tire for cars to use in the snow. They use 50 identical cars—the same make, the same color, and the same weight. The scientists have them driven over the same course at the same speed and they ask the drivers about how much control they had while driving the car. Half of the cars have new tires of an old, common style. Half of the cars have the new tires. The scientists don't tell the drivers who had the experimental tires.

In this experiment, what is the independent variable?

○ kind of tire

○ color of tire

○ opinion of the drivers

○ kind of car

Go On

15 Which of the following is **not** a property of nutrient molecules that helps determine how it gets inside the cell?

- ○ molecular size
- ○ molecular type
- ○ molecular shape
- ○ molecular charge

16 Name **two** types of technologies that help people study the surface and composition of Earth.

_____

_____

_____

_____

17 The smallest unit of an element that still retains the properties of that element is

- ○ a compound.
- ○ a molecule.
- ○ a proton.
- ○ an atom.

18 As a car is moving down the street it has considerable energy. When the brakes are applied, the energy in the system changes.

Explain **two** energy changes that happen when the driver hits the brakes.

_____

_____

_____

_____

_____

Go On

19 During warm weather when a plant is actively photosynthesizing, it will transport fluids and nutrients down the stem to the roots through which tissue?

○ xylem

○ phloem

○ cork

○ pith

20 **A student thinks that his garden will grow faster when worms are in the soil than when they are not. He wants to test this idea.**

Design an experiment he can perform to test his idea.

_____

_____

_____

_____

_____

_____

Go On

21 Photosynthesis uses sunlight to make sugar from water and carbon dioxide in the presence of chlorophyll.

What is the other product from this reaction?

○ carbon monoxide

○ oxygen

○ wastewater

○ heat

22 The chemical formula for styrene (a plastic) is $C_6H_5CH = CH_2$.

What is the number of carbon atoms in one molecule of styrene?

○ six

○ seven

○ eight

○ You can't tell from this formula.

Go On

Directions

The data below shows the properties of several elements. Use the information shown on the table to do Numbers 23 and 24.

Element	Mass	Protons	Neutrons
Hydrogen	1	1	0
Helium	4	2	2
Lithium	7	3	4
Beryllium	9	4	5
Boron	11	5	6
Carbon	12	6	6
Nitrogen	14	7	7
Oxygen	16	8	8
Fluorine	19	9	10
Neon	20	10	10

23 Write **two** conclusions that can be made from the data above.

Conclusion 1

_____

_____

_____

Conclusion 2

_____

_____

_____

Go On

24 Based on the data, which statement is **true**?

○ Hydrogen has no mass.

○ Fluorine and neon have the same number of protons.

○ Fluorine and neon have the same number of neutrons.

○ Helium, carbon, and oxygen act the same in chemical reactions.

25 Which statement about energy is **true**?

○ Energy is the ability to make objects move.

○ Energy is destroyed when you use it.

○ Energy comes in special kinds which can't be changed.

○ All of the above statements are true.

26 **White light comes from the sun as a spectrum of different wavelengths (colors). It hits the surface of Earth and everything on it, including plants.**

Which wavelength of light is **least** useful to plants?

○ red

○ orange

○ yellow

○ green

27 Look at the data below about the flight of a ball.

Time (seconds)	Height above the ground
0	0
1	2 m
2	4 m
3	6 m
4	8 m
5	9 m
6	8 m
7	6 m
8	4 m
9	2 m
10	0 m

What are **two** explanations for possible flight paths of the ball?

Explanation 1:

_____

_____

_____

Explanation 2:

_____

_____

_____

Science Assessment One—Session Two

28 Scientists and naturalists have been observing nature for hundreds of years and wondering why there are so many different kinds of organisms on Earth. Each year, there are new life forms discovered and some life forms become extinct.

Which of the following is a reasonable explanation for the diversity of life on Earth?

○ Geographic isolation creates opportunities for some organisms to reproduce and fill the available niches.

○ Vacant ecological niches stimulate competition for the resources.

○ Organisms which can compete well survive to have offspring.

○ All of the above are reasonable explanations.

29 When a substance undergoes a chemical change, like cooking an egg or neutralizing an acid, the mass of the chemicals is the same before and after the event.

Which of the following is an example of this explanation?

○ law of physical changes

○ theory of relativity

○ law of conservation of mass

○ theory of chemistry

Go On

30 A simple electric circuit including a battery, a switch, and several light bulbs is pictured below.

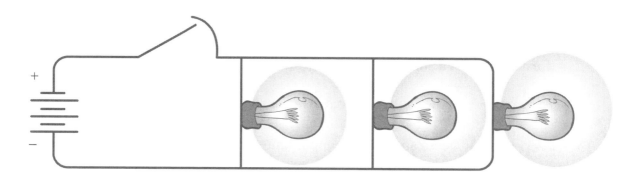

What kind of circuit is it?

○   an AC circuit

○   a battery circuit

○   a parallel circuit

○   a series circuit

31 A student noticed that he always felt itchy after eating his lunch (a peanut butter and jelly sandwich on wheat bread, a kiwi fruit, a bag of baby carrots, and a carton of milk). He hypothesized that he must be reacting to the peanut butter in his sandwich. For one week he ate the same lunch only he left out the peanut butter. Then he ate the same lunch with peanut butter for the next week. He felt itchy both weeks. He decided that his hypothesis was not supported.

Design another experiment to test his itchy lunch.

Hypothesis:

Experiment:

32 Science is practiced by many people around the world who perform experiments and investigations in different ways. They publish their results and try to compare findings.

Based on this information and what you know about science, which statement is **true**?

○ Scientists look at many designs and try to figure out which one is the right one.

○ Scientists try to keep their investigations secret so they get the answer first.

○ Scientists try to copy the work of the most famous scientists in their field.

○ Scientists will take the results from many different investigations and design new investigations.

33 When a rainbow is formed, colors with shorter wavelengths are bent more than colors with longer wavelengths.

Which color is bent the **most**?

○ red

○ green

○ yellow

○ violet

34 An experiment is being performed to test the effect of the amount of rain on the erosion of a hill.

What is the independent variable?

_____

_____

What is the dependent variable?

_____

_____

35 **Because of the amount of energy in the particles of one phase of matter, it takes the shape of the container it is in even with the lid off.**

Which phase of matter does this describe?

- ○ solid
- ○ plasma
- ○ gas
- ○ liquid

36 In a simple electrical circuit with a battery, in which direction does the energy flow?

- ○ Electrical energy flows from – to +.
- ○ Electrical energy flows from + to –.
- ○ Electrical energy can flow in either direction.
- ○ None of the above.

Go On

37 If iron metal and oxygen gas were combined in a container, how could you tell that a new compound has been formed?

- ○ The new substance won't support burning.
- ○ The new substance has a different color.
- ○ The new substance won't bend.
- ○ All of the above.

38 **An investigation resulted in a comparison of different numbers at different times.**

What would be the **best** way to display this data?

- ○ a line graph
- ○ a bar graph
- ○ a circle graph
- ○ the original data table

39 Which statement is **true** about plants and animals?

- ○ Plants and animals both use cellular respiration 24 hours a day.
- ○ Plants use cellular respiration during the night and photosynthesis during the day.
- ○ Animals use cellular respiration and plants do not.
- ○ None of the statements above are correct.

Go On

40 Look at the illustration below and the chart of the density of five different liquids.

On the lines next to the illustration, label the liquids in order from top to bottom based on their density.

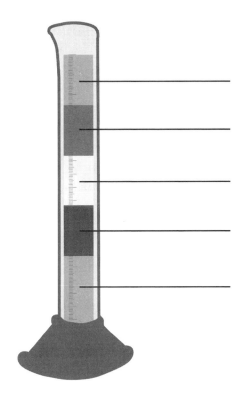

Number	Liquid	Density (g/ml)
1	Alcohol	0.81
2	Corn Syrup	1.38
3	Gasoline	0.66
4	Milk	1.03
5	Water	1.00

41 Organelles in cells, like organs in the body, have specific functions they must perform to help an organism live.

Which organelle is responsible for cellular respiration?

○ chloroplast

○ nucleus

○ cell membrane

○ mitochondrion

42 Investigators on Darwin Island in the Galapagos Islands noticed a decline in the number of small lizards from the previous year.

Write a valid scientific question that could be asked in this situation and suggest a way the question could be investigated.

Scientific question:

_____

Investigation:

_____

_____

_____

_____

_____

Go On

43 Diseases can be communicable or noncommunicable based on whether or not you can catch them from another person.

Which of the following is considered a communicable disease?

○ multiple sclerosis
○ cold sores
○ cancer
○ obesity

44 Which of the following is an example of a chemical change?

○ salt dissolving in water
○ a nail rusting
○ breaking a glass
○ shredding paper

45 Study the food chain below.

What could be happening here?

○ Pesticides can accumulate as you go up the chain.

○ More eagles could mean the extinction of the mice.

○ The amount of energy available from the sun is decreasing.

○ There is an increasing number of eagles in the ecosystem.

46 Examine the diagram of the rock layers below.

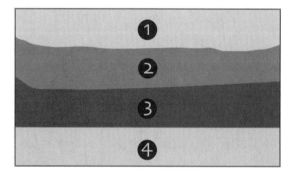

Which layer is the oldest?

○ 1

○ 2

○ 3

○ 4

Go On

47 Study the diagram of Earth's atmosphere below.

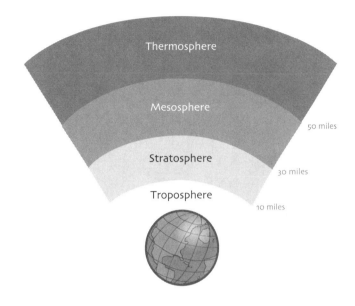

Which layer of Earth's atmosphere is the coldest?

○ thermosphere

○ mesosphere

○ stratosphere

○ troposphere

48 What happens during meiosis?

○ Two cells that are genetically different can result.

○ Two identical cells result.

○ Damaged cells are quickly replaced.

○ Half of the cells automatically die.

Go On

49 The heat from the sun heats Earth's atmosphere by which process?

○ conduction

○ convection

○ friction

○ radiation

50 In the United States, the prevailing winds come from which direction?

○ north

○ south

○ east

○ west

51 In human cells, a diploid cell has 46 chromosomes. After dividing by the process of meiosis, how many chromosomes are in the resulting cells?

○ 12

○ 23

○ 46

○ 92

Go On

52 Which of the following could be indicated by a sharp change in the general wind direction?

○ high pressure

○ low pressure

○ passing of a front

○ an inversion

53 The table below shows the percentage of specific plants in a given field sample.

Plants in a Field Sample

Plant	Percent
Wheat	62%
Annual Rye Grass	18%
Russian Thistle	11%
Clover	6%
Cinch Grass	3%

What would be the **most** informative way to display this data?

_____

Using the space below, construct a **graph** showing the data in the table. **Be sure to title your graph and indicate appropriate units.**

Go On

54 Chromosomes in the nucleus of a cell are composed of many

○ centromeres.

○ genes.

○ ribosomes.

○ vacuoles.

55 Study the drawing of the water cycle below.

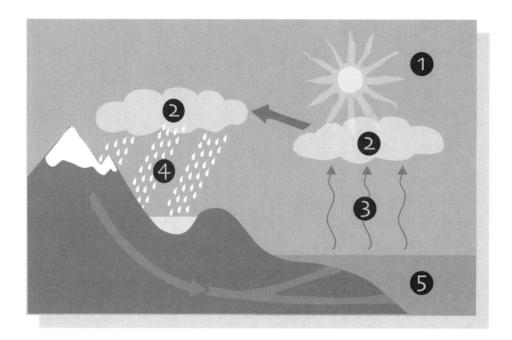

Which number indicates evaporation in this drawing?

○ 2
○ 3
○ 4
○ 5

56 Why is the ocean an ideal incubator for life?

○ Salinity almost never changes.

○ The temperature of the water seldom changes by more than a degree or two.

○ Sea water has a rich mixture of minerals and nutrients.

○ All of the above.

57 Consider the Punnett square below dealing with green pods (G) and yellow pods (g).

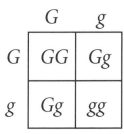

What percentage of the offspring will be yellow?

○ 25%

○ 50%

○ 75%

○ 100%

58 What is the largest planet in our solar system?

- ○ Jupiter
- ○ Saturn
- ○ Neptune
- ○ Earth

59 Which part of the diagram below shows the moon when its phase is full moon?

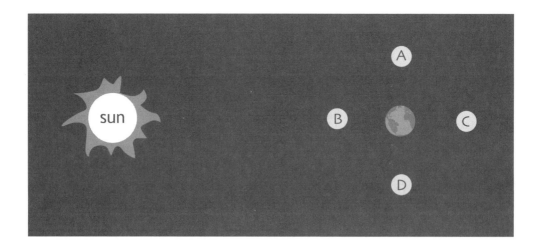

- ○ A
- ○ B
- ○ C
- ○ D

60 There are many different parts of the universe that are interesting to astronomers who wonder about its origin.

Which of the following is a spinning collection of stars?

○ a blue giant star

○ a solar system

○ a nebula

○ a galaxy

61 Which factor would **not** limit the size of a population?

○ the number of offspring in one season

○ the number of nesting sites on an island

○ the number of hiding holes in a river bank

○ the number of predators in the territory

62 Space craft and space travel have been valuable in understanding and studying the universe.

When did the first spacecraft land on the moon?

○ 1957

○ 1969

○ 1976

○ 1986

63 A student is testing the reaction of mice to colors in a maze. She uses the same three mice to run the mazes and the same reward at the end. In one maze, all of the walls are white. In the other, the walls are white except for the path to the end, which is painted red.

What is the dependent variable in this experiment?

○ the time it takes the mice to run the maze

○ the color of the walls in the mazes

○ the age of the mice

○ the type of reward

64 A control in an experiment is defined as a variable kept the same in all parts of the experiment.

Describe an experiment in which a variable is controlled.

_____

_____

_____

_____

_____

_____

_____

Go On

65 **By living in areas where other organisms inhabit, humans change the environmental conditions for those organisms.**

How might people cutting down a wooded lots to build houses affect the organisms of the area?

_____

_____

_____

_____

_____

_____

66 Give an example of a scientific idea and how it has changed over time.

_____

_____

_____

_____

_____

_____

_____

67 **Astronomy has been studied by many ancient cultures.**

Name **one** ancient culture that studied astronomy and tell where it was located.

_____

_____

_____

_____

_____

_____

_____

68 The fish fossil shown below is about 45 million years old. We still have fish that look just like this one.

Which statement is **most likely** true?

○ Fish have been the same for 45 million years.

○ The habitat of fish has changed very little in 45 million years.

○ Fish used to live in the mud.

○ Fish have evolved more than once.

***D**irections*
Study the table below which shows the distance traveled over time by an animal that is running. Use the information shown in the table to do Numbers 69 and 70.

Time (seconds)	Distance
0	0
1	2 m
2	4 m
3	6 m
4	7 m
5	8 m
6	9 m
7	10 m
8	11 m
9	12 m
10	13 m

69 Using the grid below, construct a **line graph** showing the relationship between the distance run by the animal and the time it took to run those distances. **Be sure to title your graph, label each axis, and indicate the appropriate units for each axis.**

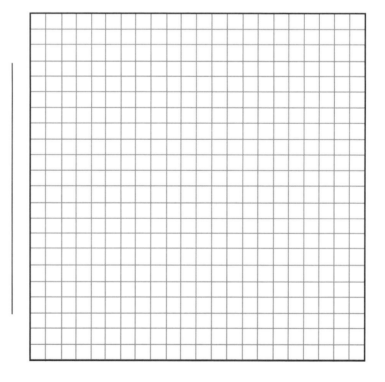

Go On

70 Where do you expect to find the animal at 12 seconds?

○ 16 m

○ 15 m

○ 14 m

○ 13 m

71 The drawing below is of cars on a roller coaster.

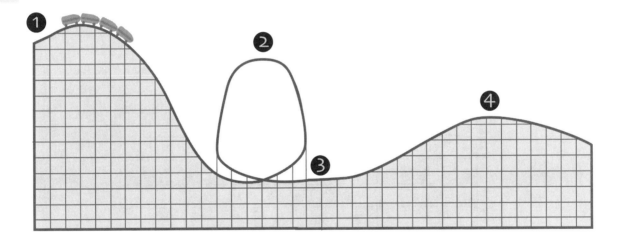

At which point on the roller coaster's path will its potential energy be the **least**?

○ Point 1

○ Point 2

○ Point 3

○ Point 4

72 When an environment is altered, some animals survive and some do not. Why?

○ Some organisms have characteristics which allow them to survive in the altered environment.

○ Some organisms can morph their bodies to adapt to the new environment.

○ Some organisms do not need to compete for resources.

○ Many organisms survive because they last longer in the altered environment.

Directions

Study the data below which shows the planetary year and diameter of the planets in our solar system. Use the information shown to do Numbers 73 and 74.

Planet	Planetary Year	Diameter
Mercury	88 days	4,878
Venus	225 days	12,112
Earth	365 days	12,756
Mars	687 days	6,790
Jupiter	11.86 years	142,796
Saturn	29.46 years	120,660
Uranus	84 years	51,200
Neptune	164.8 years	49,000

73 Make **two** predictions from this data.

Prediction 1:

_____

_____

_____

Prediction 2:

_____

_____

_____

74 Which planet do you predict would have more moons based on the data in the table?

○ Earth probably has more moons because it has life.

○ Jupiter probably has more moons because it is the largest.

○ Neptune probably has more moons because it is the furthest away.

○ Mercury probably has more moons because it has the shortest year.

75 Look at the diagram of the rock cycle below.

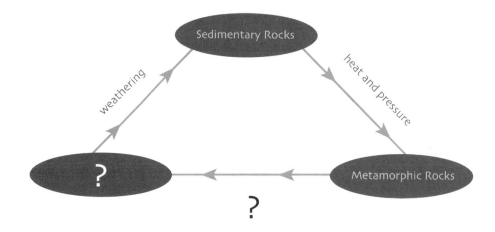

What kind of rock is missing and how is it formed?

_____

_____

_____

_____

_____

_____

_____

76 **A class is thinking about the quality of the drinking water in their school.**

Which question could be investigated with scientific design in this situation?

○ How much does the water cost?

○ Where does the water come from?

○ What kind of pipes are in the school?

○ What is the composition of the water throughout the building?

77 Which of the following is a nonrenewable source of energy?

○ solar

○ coal

○ hydroelectric

○ geothermal

78 Which model does **not** match up with its concept?

○ electron cloud ⟶ atoms

○ greenhouse effect ⟶ global warming

○ plate tectonics ⟶ volcanoes

○ energy pyramids ⟶ solar energy

Go On

79 Which of the following is a device that uses satellites to give your precise position on Earth?

- ○ telescope
- ○ microscope
- ○ GPS
- ○ Lojack

80 A molecule is the smallest part of

- ○ an atom.
- ○ a molecule.
- ○ an element.
- ○ a compound.

81 On the grid below, construct a graph showing a situation where one car is moving faster than the other starting at the same place. **Be sure to label each axis.**

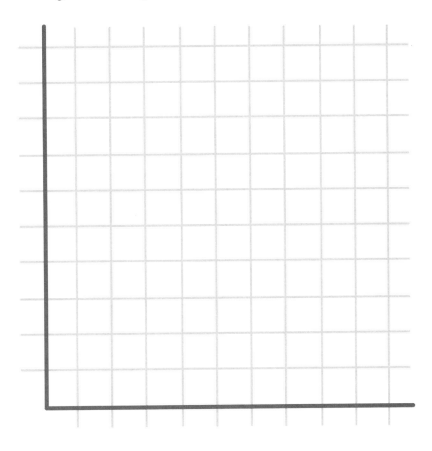

82 Choose a scientific activity and explain **one** disadvantage to the activity for the natural world.

_____

_____

_____

_____

_____

_____

83 **Solar energy can be used in many ways in our schools.**

Which of the following is a way solar energy could be converted into useful energy?

○ heating water

○ making electricity

○ stored in a chemical system to be used later

○ All of the above.

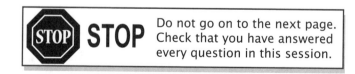
Do not go on to the next page. Check that you have answered every question in this session.

Science Assessment One—Answer Key

SESSION ONE

Identifying organs and organ systems
1 Which organ is part of the nervous system?
○ pancreas
● cerebrum
○ trachea
○ duodenum

Analysis: *The second choice is correct.* The nervous system is the brain, spinal cord, and nerves. The cerebrum is a part of the brain used for conscious thought. The first choice is incorrect. The pancreas is in the digestive system. The third choice is incorrect. The trachea is part of the respiratory system. The fourth choice is incorrect. The duodenum is part of the digestive system.

Identifying parts of an atom
2 Which particle has a negative charge and is found outside the nucleus of an atom?
● electron
○ neutron
○ prion
○ proton

Analysis: *The first choice is correct.* An electron has a negative charge and is outside the nucleus of the atom. The second choice is incorrect. Neutrons are in the nucleus and carry no charge. The third choice is incorrect. Prions are protein particles that cause diseases, such as mad cow disease. The fourth choice is incorrect. A proton is a positive charge and is found in the nucleus.

Using evidence
3 If a scientist finds evidence that does not support a hypothesis, what happens to the hypothesis?
○ The hypothesis is thrown out.
● The hypothesis is tested more.
○ A new hypothesis is automatically started.
○ The evidence is ignored because there already is a hypothesis.

Analysis: *The second choice is correct.* A hypothesis is a strong idea that has been designed to explain observations and evidence collected over a long period of time. Although a hypothesis cannot be proven to be definitely true, often, data is collected that does not support the hypothesis being tested. The first choice is incorrect. Lots of data and thinking happens before a hypothesis is rejected or replaced. The third choice is incorrect. It takes a lot of time before a hypothesis is rejected or replaced. The fourth choice is incorrect. Evidence is never ignored. It can be explained and each piece is important.

Explaining the interaction of body systems
4 **Human body systems work together to accomplish different functions for the body. The circulatory system works with many other systems.**

Name **one** system that works with the circulatory system and describe what it does.

Analysis: *Constructed-response answers may vary.* Nearly all of the body systems work together in some significant way. The answer should list a body system and describe how it works together to accomplish a bodily need. For example, one body system is the immune system. The immune system has cells that roam the circulatory system and find foreign invaders.

Explaining why Earth's surface is always changing
5 Earth is always changing.
What are **two** ways in which weathering changes the surface of Earth?

Analysis: *Constructed-response answers may vary.* Weathering can break down things like mountains or dig river gorges deeper. It can also create sediments which eventually can become new sedimentary rock.

Recording and reporting data
6 Look at the the thermometer below.

What is the temperature shown on the thermometer?
● 26 degrees C
○ 2.6 degrees C
○ 26 degrees F
○ 2.6 degrees F

Analysis: *The first choice is correct.* The thermometer measures 26 degrees Celsius (C). The second choice is incorrect. The temperature is over 20 degrees. The third choice is incorrect. The scale measures in degrees Celsius, not Fahrenheit. The fourth choice is incorrect. The scale measures in degrees Celsius, not Fahrenheit.

Science Assessment One—Answer Key

Identifying the levels of organization

7 One way of organizing how we understand the world is by looking at different levels of organization.

What level within an organism is larger than a biome?
○ molecule
● biosphere
○ organ
○ organism

Analysis: *The second choice is correct.* The biosphere is larger than a biome. The levels of organization on Earth used in science from largest to smallest are—Biosphere, Biome, Ecosystem, Community, Population, Organism, Organ System, Organ, Tissues, Cells, Organelles, Macromolecules, Molecules, and Atoms. The first choice is incorrect. Molecules are smaller than biomes. The third choice is incorrect. Organs are smaller than biomes. The fourth choice is incorrect. Organisms are smaller than biomes.

Understanding plate tectonics

8 Mountains are often formed where continental plates collide.

Which plates collided to create the Himalayas?
○ North and South America
○ Africa and Europe
○ Antarctica and Australia
● India and Asia

Analysis: *The fourth choice is correct.* The India and Asia plates collided to create the Himalayas in northern India. The first, second, and third choices are incorrect.

Evaluating data and explaining patterns seen in the past

9 There are at least two places on Earth's surface where continents seem to fit together like puzzle pieces. This was seen by Alfred Wegener in the early 20th century. His proposal was that a large supercontinent once existed.

Explain **two** pieces of additional evidence that could help support his theory.

Analysis: *Constructed-response answers may vary.* The correct answer must include two reasonable ideas that support the theory of continental drift. Some examples could include: common rock layers on different continents; common fossils on different continents; evidence that the continents are moving; earthquakes near the boundaries of continents/plates; and mountains where continents/plates have run into each other.

Evaluating data and explaining patterns seen in the past

10 The theory of plate tectonics was not widely accepted in the early 20th century because evidence of what was moving the plates could not be found. Finally, beginning in the 1950s, new technology helped find evidence that supported this theory.

What new evidence was found?
● Exploration of the ocean floor found sea floor spreading.
○ Maps of the old continents were found.
○ A scientist saw it in a dream.
○ New computer models figured it out.

Analysis: *The first choice is correct.* The exploration of the ocean floor led to the discovery of the spreading sea floor. The sea floor spreading was evidence to support the theory of plate tectonics. The second choice is incorrect. There were no people to make maps when the large supercontinents were closer together. The third choice is incorrect. Science is based in facts and evidence. The fourth choice is incorrect. Computer models first started to work on scientific ideas in the 1970s.

Identifying the physical characteristics of vertebrates

11 One physical characteristic that is used to classify vertebrates is whether or not an animal is a homeotherm. Homeotherms control their own body temperature and poikilotherms are the same temperature as their environment.

Which of the following are homeotherms?
○ fish and amphibians
○ amphibians and reptiles
○ reptiles and birds
● birds and mammals

Analysis: *The fourth choice is correct.* Both birds and mammals are homeotherms. The first choice is incorrect. Fish and amphibians are poikilotherms. The second choice is incorrect. Amphibians and reptiles are poikilotherms. The third choice is incorrect. Although birds are homeotherms, reptiles are poikilotherms.

Science Assessment One—Answer Key

Separating mixtures

12 Which property would be different enough to help separate a mixture of salt and white sand?
- ○ color
- ○ size
- ● solubility
- ○ magnetic properties

Analysis: *The third choice is correct.* Salt will dissolve in water and sand will not. The first choice is incorrect. Salt and white sand are the same color. The second choice is incorrect. Sand grains and salt crystals are very close to the same size. The fourth choice is incorrect. Neither is magnetic.

Classifying organisms

13 A student was using a tool to identify a flower he found on a school trip. Here is a portion of the tool.

```
16. _____
  a. Flower is blue ............... Blue Columbine
  b. Flower is not blue .................. go to 17
```

Which of the following is this tool?
- ○ a field guide
- ● a key
- ○ a manual
- ○ an ID card

Analysis: *The second choice is correct.* This is a part of a key. The first choice is incorrect. A field guide might have a key, but not all of them do. The third choice is incorrect. This is not a manual. The fourth choice is incorrect. ID cards are brief and not more than one page.

Identifying independent and dependent variables

14 Scientists are testing a new tire for cars to use in the snow. They use 50 identical cars—the same make, the same color, and the same weight. The scientists have them driven over the same course at the same speed and they ask the drivers about how much control they had while driving the car. Half of the cars have new tires of an old, common style. Half of the cars have the new tires. The scientists don't tell the drivers who had the experimental tires.

In this experiment, what is the independent variable?
- ● kind of tire
- ○ color of tire
- ○ opinion of the drivers
- ○ kind of car

Analysis: *The first choice is correct.* The kind of tire is the variable manipulated by the experimenter and is being tested. The second choice is incorrect. This is not a variable. All of the tires are the same color. The third choice is incorrect. The opinion of the drivers is the dependent variable. The fourth choice is incorrect. This is not a variable. All of the cars are identical.

Describing the processing of food in an organism

15 Which of the following is **not** a property of nutrient molecules that helps determine how it gets inside the cell?
- ○ molecular size
- ● molecular type
- ○ molecular shape
- ○ molecular charge

Analysis: *The second choice is correct.* A cell can't tell what type of molecule something is. The first choice is incorrect. Molecular size is important. The third choice is incorrect. Molecular shape is important. The fourth choice is incorrect. Molecular charge is important.

Using technologies

16 Name **two** types of technologies that help people study the surface and composition of Earth.

Analysis: *Constructed-response answers may vary.* The correct answer must include two reasonable technologies that help people study Earth. Correct answers could include: computers, seismographs, a compass, a GPS, sonars, radars, etc.

Elements

17 The smallest unit of an element that still retains the properties of that element is
- ○ a compound.
- ○ a molecule.
- ○ a proton.
- ● an atom.

Analysis: *The fourth choice is correct.* Elements are by definition pure substances made up of only one kind of atom. Atoms are the smallest part of an element that retain the same properties. The first choice is incorrect. Compounds are pure substances composed of two or more different kinds of atoms. The second choice is incorrect. Molecules consist of more than one atom and are not elements. The third choice is incorrect. Mixtures are composed of different substances with different atoms and molecules.

Explaining transferring and transforming of energy

18 As a car is moving down the street it has considerable energy. When the brakes are applied, the energy in the system changes.

Explain **two** energy changes that happen when the driver hits the brakes.

Analysis: *Constructed-response answers may vary.* The answers should correctly cite two energy transformations in this system. Correct answers could state that applying the brakes changes the motion of the car into heat at the brakes, sound (screeching), or motion of other objects it hits.

Science Assessment One—Answer Key

Identifying and comparing nutrient and waste transport

19 During warm weather when a plant is actively photosynthesizing, it will transport fluids and nutrients down the stem to the roots through which tissue?

○ xylem
● phloem
○ cork
○ pith

Analysis: *The second choice is correct.* The phloem moves fluids and nutrients down the stem. The first choice is incorrect. Xylem moves water and minerals up the stem. The third choice is incorrect. Cork is dead tissue which protects the stem. The fourth choice is incorrect. The pith is support tissue at the center of a stem.

Planning and designing scientific investigations

20 A student thinks that his garden will grow faster when worms are in the soil than when they are not. He wants to test this idea.

Design an experiment he can perform to test his idea.

Analysis: *Constructed-response answers may vary.* The student should divide a garden plot into two parts and plant the same plants in them, tend them the same, water them the same—everything the same—except he should buy worms and put them in one plot (independent variable), making sure the worms do not move over to the other plot. At the beginning, the student should measure the average height of the plants in the plots. After a week, he should measure them again and see if their average growth has changed (dependent variable).

Describing photosynthesis and respiration

21 Photosynthesis uses sunlight to make sugar from water and carbon dioxide in the presence of chlorophyll.

What is the other product from this reaction?

○ carbon monoxide
● oxygen
○ wastewater
○ heat

Analysis: *The second choice is correct.* Oxygen is the other product of photosynthesis. The first choice is incorrect. CO is not produced by this reaction. The third choice is incorrect. This is not a product of photosynthesis. The fourth choice is incorrect. Photosynthesis uses energy; it does not give off heat.

Explaining ratios of atoms in compounds

22 The chemical formula for styrene (a plastic) is $C_6H_5CH = CH_2$.

What is the number of carbon atoms in one molecule of styrene?

○ six
○ seven
● eight
○ You can't tell from this formula.

Analysis: *The third choice is correct.* There are 8 total carbon atoms in this chemical formula: C_6 plus C plus C is 8 total carbon atoms. The first choice is incorrect. There are 6 plus 1 plus 1 carbon atoms in this formula. The second choice is incorrect. Don't forget to add the atoms from all three parts of the formula. The fourth choice is incorrect. The formula is very precise and you call tell the number of carbon atoms.

Directions: The data below shows the properties of several elements. Use the information shown on the table to do Numbers 23 and 24.

Element	Mass	Protons	Neutrons
Hydrogen	1	1	0
Helium	4	2	2
Lithium	7	3	4
Beryllium	9	4	5
Boron	11	5	6
Carbon	12	6	6
Nitrogen	14	7	7
Oxygen	16	8	8
Fluorine	19	9	10
Neon	20	10	10

Interpreting and evaluating data

23 Write **two** conclusions that can be made from the data above.

Analysis: *Constructed-response answer may vary.* Correct answers must include two reasonable conclusions supported by this data, such as: the larger the number of protons, the bigger the mass; the number of protons plus the number of neutrons equals the mass; or every element has different numbers of protons/different mass.

Interpreting and evaluating data

24 Based on the data, which statement is **true**?

○ Hydrogen has no mass.
○ Fluorine and neon have the same number of protons.
● Fluorine and neon have the same number of neutrons.
○ Helium, carbon, and oxygen act the same in chemical reactions.

Analysis: *The third choice is correct.* Both fluorine and neon have 10 neutrons. The first choice is incorrect. The mass of hydrogen is 1. The second choice is incorrect. Fluorine has 9 protons and neon has 10. The fourth choice is incorrect. The chart does not show how helium, carbon, and oxygen act even though they each have the same number of neutrons as they have protons.

Science Assessment One—Answer Key

Recognizing forms of energy

25 Which statement about energy is **true**?
- ● Energy is the ability to make objects move.
- ○ Energy is destroyed when you use it.
- ○ Energy comes in special kinds which can't be changed.
- ○ All of the above statements are true.

Analysis: *The first choice is correct.* The ability to make objects move is the definition of energy. The second choice is incorrect. Energy cannot be created or destroyed. The third choice is incorrect. Energy can be converted from one form to another. The fourth choice is incorrect. The second and third choices are false.

Describing the colors of white light

26 White light comes from the sun as a spectrum of different wavelengths (colors). It hits the surface of Earth and everything on it, including plants.

Which wavelength of light is **least** useful to plants?
- ○ red
- ○ orange
- ○ yellow
- ● green

Analysis: *The fourth choice is correct.* When the colors of visible light are mixed, we see it as white light. When light is reflected off things, it results in the color we see. We see the plant as green because that light is being reflected off the plant. It is the least useful color to the plant. The first choice is incorrect. Red is absorbed by the plant, so we don't see this color. The second choice is incorrect. Orange is absorbed by the plant, so we don't see this color. The third choice is incorrect. Yellow is absorbed by the plant, so we don't see this color.

Describing other reasonable explanations for data or observations

27 Look at the data below about the flight of a ball.

Time (seconds)	Height above the ground
0	0
1	2 m
2	4 m
3	6 m
4	8 m
5	9 m
6	8 m
7	6 m
8	4 m
9	2 m
10	0 m

What are **two** explanations for possible flight paths of the ball?

Analysis: *Constructed-response answers may vary.* The correct answer should include two explanations that would explain the motion of the ball. For example, Explanation 1: The ball went straight up in the air and fell back down. Explanation 2: Someone hit a golf ball and it went in an arcing path from the tee to where it hit the ground again.

SESSION TWO

Describing other reasonable explanations for data or observations

28 Scientists and naturalists have been observing nature for hundreds of years and wondering why there are so many different kinds of organisms on Earth. Each year, there are new life forms discovered and some life forms become extinct.

Which of the following is a reasonable explanation for the diversity of life on Earth?
- ○ Geographic isolation creates opportunities for some organisms to reproduce and fill the available niches.
- ○ Vacant ecological niches stimulate competition for the resources.
- ○ Organisms which can compete well survive to have offspring.
- ● All of the above are reasonable explanations.

Analysis: *The fourth choice is correct.* All of the choices are reasonable explanations of the diversity of life on Earth. The first, second, and third choices are incorrect. While they are all reasonable explanations, they are not the only explanation to choose.

Science Assessment One—Answer Key

Applying the law of conservation of mass to chemical changes
29 When a substance undergoes a chemical change, like cooking an egg or neutralizing an acid, the mass of the chemicals is the same before and after the event.

Which of the following is an example of this explanation?
○ law of physical changes
○ theory of relativity
● law of conservation of mass
○ theory of chemistry

Analysis: *The third choice is correct.* This is an example of the Law of Conservation of Mass. The first choice is incorrect. There is no Law of Physical Changes. The second choice is incorrect. The Law of Relativity has to do with changes in energy and mass. The fourth choice is incorrect. There is no Theory of Chemistry.

Identifying types of circuits
30 A simple electric circuit including a battery, a switch, and several light bulbs is pictured below.

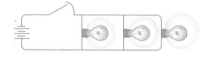

What kind of circuit is it?
○ an AC circuit
○ a battery circuit
● a parallel circuit
○ a series circuit

Analysis: *The third choice is correct.* The energy can flow through multiple paths in a parallel circuit. The first choice is incorrect. The battery makes it a DC circuit. The second choice is incorrect. There is no such thing as a battery circuit. The fourth choice is incorrect. The lights are not consecutive in a parallel circuit.

Recognizing and explaining alternative experimental designs
31 A student noticed that he always felt itchy after eating his lunch (a peanut butter and jelly sandwich on wheat bread, a kiwi fruit, a bag of baby carrots, and a carton of milk). He hypothesized that he must be reacting to the peanut butter in his sandwich. For one week, he ate the same lunch only he left out the peanut butter. Then he ate the same lunch with peanut butter for the next week. He felt itchy both weeks. He decided that his hypothesis was not supported.

Design another experiment to test his itchy lunch.

Analysis: *Constructed-response answers may vary.* The correct answer must include the following: a reasonable explanation for the observation—a hypothesis; the independent variable—something that is manipulated by the researcher; the dependent variable—the effect of the test; and controlled variables—everything else is kept the same. For example, hypothesis: The part of the student's lunch that is making him itchy is the kiwi fruit. Experiment: For one week, eat the same lunch minus the kiwi fruit and the next week eat the same lunch with the kiwi fruit. Keep everything else the same.

Recognizing and explaining alternative experimental designs
32 Science is practiced by many people around the world who perform experiments and investigations in different ways. They publish their results and try to compare findings.

Based on this information and what you know about science, which statement is **true**?
○ Scientists look at many designs and try to figure out which one is the right one.
○ Scientists try to keep their investigations secret so they get the answer first.
○ Scientists try to copy the work of the most famous scientists in their field.
● Scientists will take the results from many different investigations and design new investigations.

Analysis: *The fourth choice is correct.* Scientists will take the results from many different investigations and design new investigations. The first choice is incorrect. There is no one correct design for an investigation. The second choice is incorrect. Science is not a race to win. The third choice is incorrect. Though scientists respect the work of more experienced researchers, they will use many ideas in their design of investigations.

Science Assessment One—Answer Key

Comparing the wavelengths of colors of light

33 When a rainbow is formed, colors with shorter wavelengths are bent more than colors with longer wavelengths.

Which color is bent the **most**?
- ○ red
- ○ green
- ○ yellow
- ● violet

Analysis: *The fourth choice is correct.* The colors of the spectrum from longest to shortest are: red, orange, yellow, green, blue, indigo, and violet. Violet has the shortest wavelength of visible light. The first choice is incorrect. Red has the longest wavelength of visible light. The second choice is incorrect. Green has a longer wavelength than violet. The third choice is incorrect. Yellow has a longer wavelength than violet.

Identifying independent and dependent variables

34 An experiment is being performed to test the effect of the amount of rain on the erosion of a hill.

What is the independent variable?
What is the dependent variable?

Analysis: *Constructed-response answers may vary.* The independent variable (IV) is the variable manipulated by the experimenter or considered as the varying part that affects something else. In this case, the IV is the amount of rain. The dependent variable (DV) is the result of the experiment. In this case, the DV is the amount of erosion on the hill.

Describing the particulate model for matter

35 Because of the amount of energy in the particles of one phase of matter, it takes the shape of the container it is in even with the lid off.

Which phase of matter does this describe?
- ○ solid
- ○ plasma
- ○ gas
- ● liquid

Analysis: *The fourth choice is correct.* A liquid will take the shape of a container it is in even with the lid off. The first, second, and third choices are incorrect.

Explaining the flow of energy through a circuit

36 In a simple electrical circuit with a battery, in which direction does the energy flow?
- ● Electrical energy flows from − to +
- ○ Electrical energy flows from + to −
- ○ Electrical energy can flow in either direction.
- ○ None of the above.

Analysis: *The first choice is correct.* Electrical energy flows from − to +. The second choice is incorrect. Electrical energy flows from − to +. The third choice is incorrect. Electrical energy only flows from − to +. The fourth choice is incorrect since choice one is correct.

Describing the different properties of a compound and its atoms

37 If iron metal and oxygen gas were combined in a container, how could you tell that a new compound has been formed?
- ○ The new substance won't support burning.
- ○ The new substance has a different color.
- ○ The new substance won't bend.
- ● All of the above.

Analysis: *The fourth choice is correct.* When a compound is formed by combining two different substances, the resulting compound has new properties. All of the first three choices indicate different properties of the substance from its parts.

Recognizing different ways to communicate results

38 An investigation resulted in a comparison of different numbers at different times.

What would be the **best** way to display this data?
- ○ a line graph
- ● a bar graph
- ○ a circle graph
- ○ the original data table

Analysis: *The second choice is correct.* Bar graphs compare specific amounts at different times that are not connected. The first choice is incorrect. Line graphs show data that is constantly changing. The third choice is incorrect. Circle graphs show percents and portions of a whole. The fourth choice is incorrect. The original data table will not help the reader interpret these results.

Science Assessment One—Answer Key

Describing the relationship between photosynthesis and respiration in plants and animals

39 Which statement is **true** about plants and animals?
- ● Plants and animals both use cellular respiration 24 hours a day.
- ○ Plants use cellular respiration during the night and photosynthesis during the day.
- ○ Animals use cellular respiration and plants do not.
- ○ None of the statements above are correct.

Analysis: *The first choice is correct.* All cells must use glucose through cellular respiration to live. The second choice is incorrect. Plants use cellular respiration all the time and only use photosynthesis in the daylight. The third choice is incorrect. All organisms use cellular respiration. The fourth choice is incorrect since the first choice is correct.

Separating mixtures by density

40 Look at the illustration below and the chart of the density of five different liquids.

On the lines next to the illustration, label the liquids in order from top to bottom based on their density.

Number	Liquid	Density (g/ml)
1	Alcohol	0.81
2	Corn Syrup	1.38
3	Gasoline	0.66
4	Milk	1.03
5	Water	1.00

Analysis: The correct answers should place the liquids in this order from top to bottom:
3 or Gasoline
1 or Alcohol
5 or Water
4 or Milk
2 or Corn Syrup

Identifying cellular organelles and their function

41 Organelles in cells, like organs in the body, have specific functions they must perform to help an organism live.

Which organelle is responsible for cellular respiration?
- ○ chloroplast
- ○ nucleus
- ○ cell membrane
- ● mitochondrion

Analysis: *The fourth choice is correct.* The mitochondrion is the site of cellular respiration. The first choice is incorrect. The chloroplast is the site of photosynthesis. The second choice is incorrect. The nucleus contains the genetic material. The third choice is incorrect. The cell membrane controls transport into and out of cells.

Identifying possible scientific questions

42 Investigators on Darwin Island in the Galapagos Islands noticed a decline in the number of small lizards from the previous year.

Write a valid scientific question that could be asked in this situation and suggest a way the question could be investigated.

Analysis: *Constructed-response answers may vary.* The correct answer must include the following: The question should be a reasonable, testable, and repeatable question, and the investigation should reasonably investigate the question. For example, Scientific question: How does rainfall affect the number of small lizards on Darwin Island? Investigation: Investigators could count the number of small lizards that inhabit a specific area of Darwin Island while recording the amount of rainfall over the same period. This could take several years.

Classifying diseases as communicable and noncommunicable

43 Diseases can be communicable or noncommunicable based on whether or not you can catch them from another person.

Which of the following is considered a communicable disease?
- ○ multiple sclerosis
- ● cold sores
- ○ cancer
- ○ obesity

Analysis: *The second choice is correct.* Communicable means you can catch the disease from another person. Cold sores can be passed from person to person. The first choice is incorrect. Multiple sclerosis is not passed from person to person. The third choice is incorrect. Cancer is not passed among people. The fourth choice is incorrect. Obesity is not passed from person to person.

Distinguishing between physical and chemical changes

44 Which of the following is an example of a chemical change?
- ○ salt dissolving in water
- ● a nail rusting
- ○ breaking a glass
- ○ shredding paper

Analysis: *The second choice is correct.* A chemical change occurs when something is changed and changes its basic properties. The iron in the nail becomes iron oxide, which has different properties. The first choice is incorrect. The salt and water keep their basic properties. The third choice is incorrect. The broken glass is still glass. The fourth choice is incorrect. The shredded paper is still paper.

Science Assessment One—Answer Key

Examining the flow of energy in an ecosystem

45 Study the food chain below.

What could be happening here?
- ● Pesticides can accumulate as you go up the chain.
- ○ More eagles could mean the extinction of the mice.
- ○ The amount of energy available from the sun is decreasing.
- ○ There is an increasing number of eagles in the ecosystem.

Analysis: *The first choice is correct.* Biological magnification can pass poisons up the food chain. The second choice is incorrect. More eagles can only be supported by increasing the numbers of mice in this food chain. The third choice is incorrect. The amount of sunlight is not measured in a food chain. The fourth choice is incorrect. The number of organisms decreases as you go up the chain.

Interpreting rock layers

46 Examine the diagram of rock layers below.

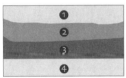

Which layer is the oldest?
- ○ 1
- ○ 2
- ○ 3
- ● 4

Analysis: *The fourth choice is correct.* According to the concept of superpositioning, younger layers are generally formed on top of older layers. Layer 4 is buried under all of the others. The first choice is incorrect. Layer 1 is above layer 4. The second choice is incorrect. Layer 2 is above layer 4. The third choice is incorrect. Layer 3 is above layer 4.

Identifying layers of the atmosphere

47 Study the diagram of Earth's atmosphere below.

Which layer of Earth's atmosphere is the coldest?
- ○ thermosphere
- ● mesosphere
- ○ stratosphere
- ○ troposphere

Analysis: *The second choice is correct.* Generally, it gets colder as you get higher in the troposphere and the mesosphere. The opposite is true in the other layers. The mesosphere layer is very cold. The first choice is incorrect. The thermosphere is the upper-most layer and closest to space; it is the hottest. The third choice is incorrect. Although the stratosphere doesn't get much above freezing, it is not as cold as the mesosphere layer. The fourth choice is incorrect. The troposphere is the layer we live in.

Differentiating between mitosis and meiosis

48 What happens during meiosis?
- ● Two cells that are genetically different can result.
- ○ Two identical cells result.
- ○ Damaged cells are quickly replaced.
- ○ Half of the cells automatically die.

Analysis: *The first choice is correct.* When the cell splits, it takes half of the chromosomes which can be different from each other. The second choice is incorrect. Meiosis results in cells that are genetically different. The third choice is incorrect. Replacing damaged cells is one function of mitosis. The fourth choice is incorrect. Half of the cells do not automatically die.

Explaining the heating of Earth by the sun

49 The heat from the sun heats Earth's atmosphere by which process?
- ○ conduction
- ● convection
- ○ friction
- ○ radiation

Analysis: *The second choice is correct.* Convective heating moves the heat through the atmosphere as it circulates the air. The first choice is incorrect. Conduction requires contact with the object heating it. The third choice is incorrect. Friction is from mechanical contact. The fourth choice is incorrect. The heat from the sun travels through space as radiant energy that strikes the solid Earth that heats up the air.

Interpreting weather data

50 In the United States, the prevailing winds come from which direction?
- ○ north
- ○ south
- ○ east
- ● west

Analysis: *The fourth choice is correct.* The prevailing winds in the U.S. are from the west. The first, second, and third choices are incorrect.

Relating the numbers of chromosomes to mitosis and meiosis

51 In human cells, a diploid cell has 46 chromosomes. After dividing by the process of meiosis, how many chromosomes are in the resulting cells?
- ○ 12
- ● 23
- ○ 46
- ○ 92

Analysis: *The second choice is correct.* Mitosis is a body process that starts and ends with diploid cells. Meiosis results in haploid cells; 23 is half of 46. The first choice is incorrect. Twelve is not quite half of half of 46. The third choice is incorrect. Forty-six is the diploid number. The fourth choice is incorrect. The number 92 would be double the diploid number.

Science Assessment One—Answer Key

Identifying causes of changes in the weather

52 Which of the following could be indicated by a sharp change in the general wind direction?
○ high pressure
○ low pressure
● passing of a front
○ an inversion

Analysis: *The third choice is correct.* Drastic wind changes usually mean that a front has just passed. The first choice is incorrect. The pressure cell could change the wind direction if it was moving through with a front. The second choice is incorrect. The pressure cell could change the wind direction if it was moving through with a front. The fourth choice is incorrect. An inversion creates stable air.

Recognizing different ways to communicate results

53 The table below shows the percentage of specific plants in a given field sample.

Plants in a Field Sample

Plant	Percent
Wheat	62%
Annual Rye Grass	18%
Russian Thistle	11%
Clover	6%
Cinch Grass	3%

What would be the **most** informative way to display this data?

Using the space below, construct a graph showing the data in the table. Be sure to title your graph and indicate appropriate units.

Analysis: *Constructed-response answers may vary.* The best way to display this data is with a circle graph.

Describing the roles of chromosomes and genes in heredity

54 Chromosomes in the nucleus of a cell are composed of many
○ centromeres.
● genes.
○ ribosomes.
○ vacuoles.

Analysis: *The second choice is correct.* Genes compose the chromosomes in the nucleus of a cell. The first choice is incorrect. A chromosome has one centromere; centromeres assist in cell reproduction. The third choice is incorrect. Ribosomes are not on the chromosomes. The fourth choice is incorrect. Vacuoles are not on the chromosomes.

SESSION THREE

Explaining the water cycle

55 Study the drawing of the water cycle below.

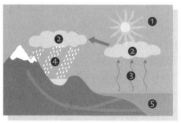

Which number indicates evaporation in this drawing?
○ 2
● 3
○ 4
○ 5

Analysis: *The second choice is correct.* Number 3 indicates evaporation. The first choice is incorrect. Number 2 is condensation in the clouds. The third choice is incorrect. Number 4 is precipitation. The fourth choice is incorrect. Number 5 shows liquid water in the ocean.

Understanding the composition and characteristics of oceans

56 Why is the ocean an ideal incubator for life?
○ Salinity almost never changes.
○ The temperature of the water seldom changes by more than a degree or two.
○ Sea water has a rich mixture of minerals and nutrients.
● All of the above.

Analysis: *The fourth choice is correct.* The ocean is the most stable environment on Earth. The first, second, and third choices are all reasons why the ocean is an ideal incubator for life.

Science Assessment One—Answer Key

Inferring the traits of offspring based on the genes of the parents

57 Consider the Punnet square below dealing with green pods (G) and yellow pods (g).

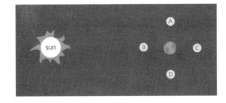

What percentage of the offspring will be yellow?
- ● 25%
- ○ 50%
- ○ 75%
- ○ 100%

Analysis: *The first choice is correct.* In a case like this, green is dominant to the yellow color in seed pods. Both GG and Gg genotypes would produce green pods. The genotype gg is yellow. The second choice is incorrect. Gg and GG are both green. The third choice is incorrect. Three out of four are green. The fourth choice is incorrect. Both GG and Gg genotypes would produce green pods, so 100% of the offspring could not be yellow.

Describing the solar system

58 What is the largest planet in our solar system?
- ● Jupiter
- ○ Saturn
- ○ Neptune
- ○ Earth

Analysis: *The first choice is correct.* Jupiter is a gaseous giant planet. The second choice is incorrect. Saturn is also a gaseous giant but not as large as Jupiter. The third choice is incorrect. Neptune is the third largest planet and only about 1/3 of the size of Jupiter. The fourth choice is incorrect. Earth is one of the smaller planets in our solar system.

Understanding the phases of the moon and tides

59 Which part of the diagram below shows the moon when its phase is full moon?

- ○ A
- ○ B
- ● C
- ○ D

Analysis: *The third choice is correct.* The moon is full in postion C of the diagram. The first choice is incorrect. The moon is about half full in this position. The second choice is incorrect. This is a new moon. The fourth choice is incorrect. The moon is about half full in this position.

Describing the components of the universe

60 There are many different parts of the universe that are interesting to astronomers who wonder about its origin.

Which of the following is a spinning collection of stars?
- ○ a blue giant star
- ○ a solar system
- ○ a nebula
- ● a galaxy

Analysis: *The fourth choice is correct.* A galaxy is a collection of stars spinning around a central point. The first choice is incorrect. A blue giant is just one star. The second choice is incorrect. A solar system usually consists of a star and its planets. The third choice is incorrect. A nebula is a cloud of interstellar dust.

Describing environmental factors that limit population size

61 Which factor would **not** limit the size of a population?
- ● the number of offspring in one season
- ○ the number of nesting sites on an island
- ○ the number of hiding holes in a river bank
- ○ the number of predators in the territory

Analysis: *The first choice is correct.* The number of offspring in one season would likely not be a limiting factor. The second choice is incorrect. Fewer nests means fewer offspring. The third choice is incorrect. Fewer places to hide means more predation and a smaller population. The fourth choice is incorrect. More predators means more predation and a smaller population.

Understanding technologies needed for the exploration of space

62 Space craft and space travel have been valuable in understanding and studying the universe.

When did the fist spacecraft land on the moon?
- ○ 1957
- ● 1969
- ○ 1976
- ○ 1986

Analysis: *The second choice is correct.* Neil Armstrong first stepped on the moon July 16, 1969. The first choice is incorrect. The first satellite, Sputnik, went up in 1957. The third choice is incorrect. This was the bicentennial of the United States. The fourth choice is incorrect. This was the year of the space shuttle *Challenger* disaster.

Science Assessment One—Answer Key

Identifying independent and dependent variables
63 A student is testing the reaction of mice to colors in a maze. She uses the same three mice to run the mazes and the same reward at the end. In one maze, all of the walls are white. In the other, the walls are white except for the path to the end, which is painted red.

What is the dependent variable in this experiment?
- ● the time it takes the mice to run the maze
- ○ the color of the walls in the mazes
- ○ the age of the mice
- ○ the type of reward

Analysis: *The first choice is correct.* The time it takes for the mice to run the maze is the dependent variable because it is the result of varying the independent variable. The second choice is incorrect. The color of the walls in the maze is the independent variable because it is manipulated by the student. The third choice is incorrect. The age of the mice is the same for all trials. The fourth choice is incorrect. The type of reward does not change.

Identifying controlled factors in an investigation
64 A control in an experiment is defined as a variable kept the same in all parts of the experiment.

Describe an experiment in which a variable is controlled.

Analysis: *Constructed-response answers may vary.* Correct answers will describe holding the controlled variables all the same in an experiment and testing only one variable, i.e. having only one thing changing in the trials.

Describing the impact of humans on the environment
65 By living in areas where other organisms inhabit, humans change the environmental conditions for those organisms.

How might people cutting down a wooded lots to build houses affect the organisms of the area?

Analysis: *Constructed-response answers may vary.* Humans can impact an environment in many ways. The effects cited by the student could include fewer organisms, different organisms, destruction of the habitat for animals, less food and living space for organisms, or changing the conditions so different organisms, like weeds, can move in.

Describing why scientific knowledge changes over time
66 Give an example of a scientific idea and how it has changed over time.

Analysis: *Constructed-response answers may vary.* A scientific idea, like the atomic theory, should be cited and one way the idea has changed over time should be noted. For example, the scientific idea of an atom has changed over time. At one time, we thought of atoms like balls, now we think of them like clouds of energy.

Recognizing multicultural contributions to science
67 Astronomy has been studied by many ancient cultures.

Name **one** ancient culture that studied astronomy and tell where it was located.

Analysis: *Constructed-response answers may vary.* A valid ancient culture should be named (Egypt, Greece, Maya) and that culture's correct location (Africa, Asia, Central America).

Comparing evidence of past life
68 The fish fossil shown below is about 45 million years old. We still have fish that look just like this one.

Which statement is **most likely** true?
- ○ Fish have been the same for 45 million years.
- ● The habitat of fish has changed very little in 45 million years.
- ○ Fish used to live in the mud.
- ○ Fish have evolved more than once.

Analysis: *The second choice is correct.* There has been very little basic change in fish over geologic time since the ocean has stayed very close to the same. The first choice is incorrect. We know there are different kinds of fish now. The third choice is incorrect. We don't know much about how these fish lived, but we do know this one died and fell into the mud and was fossilized. The fourth choice is incorrect. We have fossilized fish throughout the fossil record since they originally appeared millions of years ago.

Science Assessment One—Answer Key

Directions: Study the table below which shows the distance traveled over time by an animal that is running. Use the information shown in the table to do Numbers 69 and 70.

Time (seconds)	Distance
0	0
1	2 m
2	4 m
3	6 m
4	7 m
5	8 m
6	9 m
7	10 m
8	11 m
9	12 m
10	13 m

Constructing visual methods to summarize data

69 Using the grid below, construct a **line graph** showing the relationship between the distance run by the animal and the time it took to run those distances. **Be sure to title your graph, label each axis, and indicate the appropriate units for each axis.**

Analysis: *Constructed-response answers may vary.* The correct answer must include the following: a title, the vertical axis should be labeled "Time," the horizontal axis should be labeled "Distance," and the line given should accurately show the data as presented.

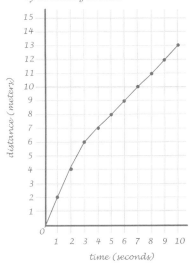

Constructing visual methods to summarize data

70 Where do you expect to find the animal at 12 seconds?
○ 16 m
● 15 m
○ 14 m
○ 13 m

Analysis: *The second choice is correct.* The data shows that the animal is increasing its distance one meter for each time point. If this is the case, then at 12 seconds, the animal would have moved a distance of 15 m. The first choice is incorrect. A distance of 16 m is more than one meter per time point. The third choice is incorrect. A distance of 14 m is less than one meter per time point. The fourth choice is incorrect. A distance of 13 m would happen if the animal stopped and there is no information to show this.

Comparing potential and kinetic energy

71 The drawing below is of cars on a roller coaster.

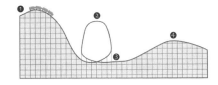

At which point on the roller coaster's path will its potential energy be the **least**?
○ Point 1
○ Point 2
● Point 3
○ Point 4

Analysis: *The third choice is correct.* Point 3 is lowest point on the drawing. If the roller coaster started here, it would probably not move. It probably has the highest kinetic energy here if it started at Point 1 and converted all of that potential energy to kinetic energy. The first choice is incorrect. The cars are the highest at Point 1. The second choice is incorrect. Point 3 is lower than Point 2. The fourth choice is incorrect. Point 3 is lower than Point 4.

Evaluating the potential of an organism with specific traits to survive and reproduce

72 When an environment is altered, some animals survive and some do not. Why?
● Some organisms have characteristics which allow them to survive in the altered environment.
○ Some organisms can morph their bodies to adapt to the new environment.
○ Some organisms do not need to compete for resources.
○ Many organisms survive because they last longer in the altered environment.

Analysis: *The first choice is correct.* Sometimes organisms already have characteristics that help them or they have the ability to change their behavior so they can survive to reproduce. The second choice is incorrect. Once an organism has developed, it can no longer adapt its characteristics, only survive to reproduce or not. The third choice is incorrect. All organisms are in competition for resources in nature. The fourth choice is incorrect. Only the organisms which can live long enough to have generations of young will survive. It is not about individuals but populations of organisms.

Science Assessment One—Answer Key

Directions: Study the data table below which shows the planetary year and diameter of the planets in our solar system. Use the information shown to do Numbers 73 and 74.

Planet	Planetary Year	Diameter
Mercury	88 days	4,878
Venus	225 days	12,112
Earth	365 days	12,756
Mars	687 days	6,790
Jupiter	11.86 years	142,796
Saturn	29.46 years	120,660
Uranus	84 years	51,200
Neptune	164.8 years	49,000

Making predictions
73 Make **two** predictions from this data.
Analysis: *Constructed-response answers may vary.* Prediction 1: Smaller planets have less gravity. Prediction 2: Planets with longer years are further from the sun and, therefore, colder.

Making predictions
74 Which planet do you predict would have more moons based on the data in the table?
○ Earth probably has more moons because it has life.
● Jupiter probably has more moons because it is the largest.
○ Neptune probably has more moons because it is the furthest away.
○ Mercury probably has more moons because it has the shortest year.
Analysis: *The second choice is correct.* Larger planets have more gravity and can capture more moons. The first choice is incorrect. The effect of the amount of life on a planet's number of moons is not included in the table. The third choice is incorrect. Distance from the sun does not make sense as a reason why a planet captures more moons. The fourth choice is incorrect. The planetary year does not make sense as a reason to predict the most moons.

Understanding the rock cycle
75 Look at the diagram of the rock cycle below.

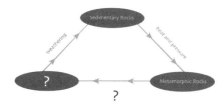

What kind of rock is missing and how is it formed?
Analysis: *Constructed-response answers may vary.* The correct type of rock is igneous rock, which is formed by metamorphic rock melting into magma, then cooling and hardening into igneous rock.

Identifying possible scientific questions
76 A class is thinking about the quality of the drinking water in their school.
Which question could be investigated with scientific design in this situation?
○ How much does the water cost?
○ Where does the water come from?
○ What kind of pipes are in the school?
● What is the composition of the water throughout the building?
Analysis: *The fourth choice is correct.* The composition of the water could be designed and investigated in a controlled scientific manner. The first choice is incorrect. How much the water costs is a simple research question. The second choice is incorrect. Where the water comes from is a simple research question. The third choice is incorrect. The type of pipes in the school is a simple research question.

Understanding the differences between renewable and nonrenewable energy
77 Which of the following is a nonrenewable source of energy?
○ solar
● coal
○ hydroelectric
○ geothermal
Analysis: *The second choice is correct.* Coal is a nonrenewable source of energy. It takes so long to make coal that it cannot be replaced in our lifetime. The first choice is incorrect. The sun is a source of energy that is seemingly inexhaustible as long as we don't block the sky. The third choice is incorrect. When water moves through a generator, it makes electricity. This is renewable as long as we have gravity and can find water. The fourth choice is incorrect. Geothermal energy is a renewable resource that captures the power of steam from Earth to generate electricity.

Describing appropriate models for scientific processes
78 Which model does **not** match up with its concept?
○ electron cloud ⟶ atoms
○ greenhouse effect ⟶ global warming
○ plate tectonics ⟶ volcanoes
● energy pyramids ⟶ solar energy
Analysis: *The fourth choice is correct.* Energy pyramids explain energy in ecological systems. The first choice is incorrect. An electron cloud is a model that explains atoms. The second choice is incorrect. The greenhouse effect is a model that helps explain global warming. The third choice is incorrect. Plate tectonics explain how volcanoes form and predict where they will be.

Science Assessment One—Answer Key

Using technologies
79 Which of the following is a device that uses satellites to give your precise position on Earth?
○ telescope
○ microscope
● GPS
○ Lojack

Analysis: *The third choice is correct.* A GPS—Global Positioning System—uses signals from several satellites to give you a precise reading of your location on Earth. The first choice is incorrect. Telescopes are for studying distant objects. The second choice is incorrect. Microscopes are used to study tiny objects and organisms. The fourth choice is incorrect. A Lojack is a radio device used to recover stolen cars.

Identifying that the smallest unit of a compound is a molecule
80 A molecule is the smallest part of
○ an atom.
○ a molecule.
○ an element.
● a compound.

Analysis: *The fourth choice is correct.* A compound is composed of molecules. The first choice is incorrect. An atom is the smallest part of an element. The second choice is incorrect. Molecules are made of atoms. The third choice is incorrect. The smallest part of an element is an atom.

Using measurements for objects moving in a straight line
81 On the grid below, construct a graph showing a situation where one car is moving faster than the other starting at the same place. **Be sure to label each axis.**
Analysis:

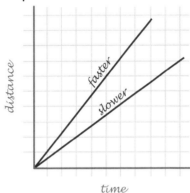

Explaining that human activity can have positive and negative effects on the natural world
82 Choose a scientific activity and explain **one** disadvantage to the activity for the natural world.

Analysis: *Constructed-response answers may vary.* A correct answer will name a scientific activity, like improving mining or improving gas mileage for vehicles, and cite one reasonable disadvantage the activity has for the natural world, like increased mining.

Identifying energy transformations in a system
83 Solar energy can be used in many ways in our schools.
Which of the following is a way solar energy could be converted into useful energy?
○ heating water
○ making electricity
○ stored in a chemical system to be used later
● All of the above.

Analysis: *The fourth choice is correct.* Solar energy can be converted into many forms. The first, second, and third choices are all good ways to use solar energy.

Science Assessment One—Session One: Correlation Chart

Use this chart to identify areas for improvement for individual students or for the class as a whole. For example, enter students' names in the left-hand column. When a student misses a question, place an "X" in the corresponding box. A column with a large number of "Xs" shows that the class needs more practice with that particular objective.

Correlation	3.2.a	2.5.a	1.3.b	3.2.b	4.3.a	1.2.a	3.3.a	4.4.a	1.4.a	1.4.a	3.1.a	2.2.a	3.1.b	1.1.b	3.4.a	1.2.b	2.5.b	2.8.b	3.4.b	1.1.a	3.5.a	2.6.a	1.3.a	1.3.a	2.8.a	2.10.a	1.5.a
Question	1	2	3	4	5	6	7	8	9	10	11	12	13	14	15	16	17	18	19	20	21	22	23	24	25	26	27

CSAP Science for Grade 8 — Science Assessment One—Correlation Chart

Science Assessment One—Session Two: Correlation Chart

Correlation	1.5.a	2.3.c	2.9.b	1.5.b	1.5.b	2.10.b	1.1.b	2.1.a	2.9.a	2.6.b	1.6.a	3.5.b	2.2.b	3.6.a	1.1.c	3.7.a	2.3.a	3.8.a	4.6.a	4.7.a	3.9.a	4.8.a	4.9.a	3.9.b	4.10.a	1.6.a	3.10.a
Question	28	29	30	31	32	33	34	35	36	37	38	39	40	41	42	43	44	45	46	47	48	49	50	51	52	53	54

Students' names

© Englefield & Associates, Inc. Copying is Prohibited Student Self-Study Workbook 315

Science Assessment One—Session Three: Correlation Chart

Correlation	4.11.a	4.12.a	3.10.b	4.13.a	4.14.a	4.15.a	3.11.a	4.16.a	1.1.b	5.1.a	3.11.b	5.2.a	5.3.a	3.12.a	1.2.c	1.2.c	2.7.c	3.13.a	1.3.c	1.3.c	4.1.a	1.1.c	4.2.a	5.4.b	1.2.b	2.6.d	2.7.a	5.5.a	2.8.c
Question	55	56	57	58	59	60	61	62	63	64	65	66	67	68	69	70	71	72	73	74	75	76	77	78	79	80	81	82	83

Science Assessment Two

Directions for Science Assessment Two

This Grade 8 CSAP Science Assessment (Session One, Session Two, and Session Three) has multiple-choice and constructed-response questions. There are several important things to remember as you take the Science CSAP:

- Only No. 2 pencils may be used on any part of the test materials. This includes the front cover.
- During testing, no one is allowed to have electronic communication devices in the testing room.
- Read each multiple-choice question carefully. Think about what is being asked. Then fill in one answer bubble to mark your answer.
- If you do not know the answer to a multiple-choice question, skip it and go on.
- You **may** use the space provided in the test book. You **must** show all of your work in the space or on the lines provided.
- For constructed-response questions, write your response clearly and neatly on the lines provided.
- If you finish a session early, you may go back and check over your work on that session only. You may be allowed to read, but you are **not** allowed to write.

1 Look at the diagram of the rock cycle below.

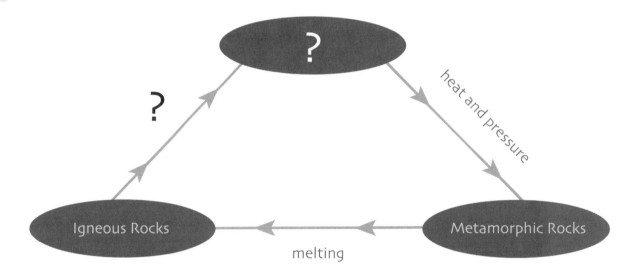

What kind of rock is missing and how is it formed?

_____

_____

_____

_____

_____

_____

2 Many scientific ideas have had both advantages and disadvantages for the world. One example is nuclear power.

Name **one** advantage and **one** disadvantage for nuclear power.

Advantage:

_____

_____

Disadvantage:

_____

_____

3 Which property would be **best** to separate a mixture of iron filings and coal dust?

○ color

○ size

○ solubility

○ magnetism

4 When may a model have an advantage over direct observations?

○ when processes are very dangerous
○ when processes have only happened once
○ when observations might take too long
○ All of the above.

5 If an environment changed from very dry to very wet, which organism would survive and reproduce?

○ one that was more generally adapted to many different habitats
○ one that was well adapted to the dry environment
○ one that could change its physical characteristics
○ one that reproduces asexually by cloning

6 **A class is thinking about the food in its school's cafeteria.**

Which question could be investigated with scientific design in this situation?

○ What is the price of a student lunch?
○ How does the amount of fat in the food affect the amount of food sold?
○ Who cooks the food?
○ Where does the food come from?

Go On

7 Which model matches up with its concept?

- ○ electron cloud ⟶ weather
- ○ greenhouse effect ⟶ plant genetics
- ○ plate tectonics ⟶ volcanoes
- ○ energy pyramids ⟶ solar energy

8 The picture below is of a fossilized brachiopod from millions of years ago. Fossilized brachiopods are found in limestone all over the world. There are still living brachiopods in the oceans of the world.

To become a fossil an organism must

- ○ fall into the mud and get covered up by more mud.
- ○ live in the ocean.
- ○ have a shell.
- ○ have been around at least 100 million years ago.

9 A student noticed that she always felt sleepy after lunch. She hypothesized that the starches and sugars in her lunch were making her sleepy. She designed an experiment in which she ate no bread, crackers, or sugars with her lunch for a week and checked how she felt. Then she went back to her old lunch habits and compared how she felt in the two hours after lunch. She decided her hypothesis was supported.

Design another experiment to test her sleepiness.

Hypothesis:

_____

_____

Experiment:

_____

_____

_____

_____

_____

_____

Go On

10 An investigator is observing animals in their natural habitat. He hypothesizes that they act very differently when he is there than when he is not there.

Which experimental design will **not** work to test his hypothesis?

○ Set up a video camera in a tree to observe then compare how the animals act to how they acted when he was present.

○ Sit quietly and do not take notes then compare how the animals act to how they acted when he was present and taking notes.

○ Make a blind and hide from the animals then compare how the animals act to how they acted when he was present and not hidden.

○ Perform his observations from a mile away using a powerful telescope then compare how the animals act to how they acted when he was present.

11 Which of the following is a cooperative study between cultures?

○ the space station
○ studies of global warming
○ automotive engineering
○ All of the above are multicultural efforts.

12 When changes in the environment occur, the organisms who live there must respond.

Which of the following is a **likely** response to pollution in an environment?

○ Organisms breed more often.
○ Many organisms will die.
○ Organisms change their eating habits.
○ Organisms clean their habitat.

Go On

13 **A scientist thinks that a new drug treatment that he has invented will cure a disease. He wants to test this idea.**

Design an experiment he can perform to test his idea.

_____

_____

_____

_____

_____

_____

14 Which scientific idea has stayed the same over time?

○ the atomic theory

○ the theory of natural selection

○ germ theory

○ None of the above

15 **Sometimes changes in an environment affect the number of organisms in a population.**

How could the amount of food and nesting sites limit the size of a population?

_____

_____

_____

_____

_____

16 In a house circuit, what direction does the electrical energy flow?

○ Electrical energy flows from – to +.

○ Electrical energy flows from + to –.

○ Electrical energy alternates directions.

○ None of the above

17 Name **two** types of technologies that help people study the ocean.

Technology 1:

_____

_____

Technology 2:

_____

_____

18 When an experiment is completed, what is the **most** important thing an investigator can do to be sure of his or her results?

○ publish the results

○ repeat the experiment

○ change the variables and start over

○ remove the controls and repeat the experiment

19 Consider the Punnett square below dealing with round seeds (R) and wrinkled seeds (r).

	R	r
R	RR	Rr
r	Rr	rr

What percentage of the offspring will be round?

○ 25%

○ 50%

○ 75%

○ 100%

20 In an experiment testing dog food, a student made sure he held all of the conditions the same except the kind of food he served the dog.

What are **two** likely control variables in this investigation?

_____

_____

_____

_____

_____

Go On

21 When you lift a pendulum all the way to the top, energy is stored in the ball as potential energy. This energy is converted to motion as the ball falls and is stored again when the ball swings up again.

Why won't the pendulum go on forever?

○ Some of the energy is destroyed by friction.

○ Some of the energy lets you see the pendulum swing.

○ Some of the energy is transformed into heat by friction.

○ Some of the energy is used to make the ball heavier at the bottom of the swing.

22 A class is testing paper towels to see which one is the strongest when wet. They are using several brands of paper towels. In each case, they wet the towel and then apply weights in 10 gram increments to see when the towel breaks.

What is the independent variable in this experiment?

○ source of the towels

○ weight when the towel breaks

○ amount of water used to wet the towels

○ brand of towel

23 In a forest fire, there are many transformations of energy.

What are **two** energy changes that would take place during a forest fire?

_____

_____

_____

_____

_____

24 In the electrical system of a house, when a light bulb goes out, the others will still work.

What kind of circuit must this be?

○ a DC circuit

○ a parallel circuit

○ a series circuit

○ an open circuit

25 In human cells, a diploid cell has 46 chromosomes. After dividing by the process of mitosis, how many chromosomes are in the resulting cells?

○ 12

○ 23

○ 46

○ 92

26 Which tool can be used to determine the composition of a star?

○ a telescope

○ a spectroscope

○ a microscope

○ an electronic scale

27 What is the mass of the ball on the scale shown below?

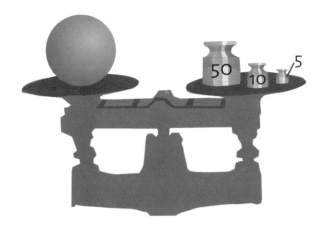

- ○ 65 grams
- ○ 65 pounds
- ○ 65 kilograms
- ○ 6.5 grams

Science Assessment Two—Session Two

28 In mitosis,

○ two cells that are genetically different can result.

○ two identical cells result.

○ sex cells are produced.

○ half of the cells automatically die.

29 **Weight and mass are different.**

What is being measured when you measure weight?

○ number of atoms

○ amount of particles

○ force of planetary spin on an object

○ force of gravity on an object

30 The sketch below is a very different version of a food pyramid.

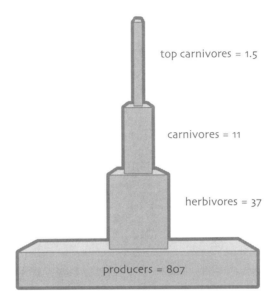

What is being measured in this food pyramid?

○ numbers of organisms

○ weight of each organism

○ biomass in the system

○ sunlight received by each level

31 There are many different parts of the universe that are interesting to astronomers who wonder about its origin.

Which of the following is an area where stars develop?

○ a blue giant star

○ a solar system

○ a nebula

○ a galaxy

32 Which part of the diagram below shows the moon when its phase is new moon?

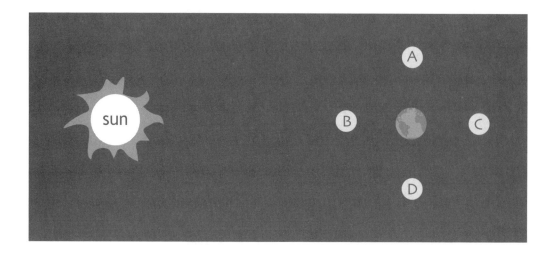

○ A
○ B
○ C
○ D

33 A gardener would like to know what to plant on the west side of her greenhouse.

Write a valid scientific question that could be asked in this situation and suggest a way the gardener might investigate the question.

The question:

The investigation:

34 White light comes from the sun as a spectrum of different wavelengths (colors). It is affected by the atmosphere as it passes through.

Which is an effect of this interaction between sunlight and the atmosphere?

○ red sunrises

○ orange sunsets

○ blue sky

○ All of the above.

35 Which planet in our solar system is the smallest?

○ Mercury

○ Venus

○ Uranus

○ Earth

36 Which of the following is a producer in a terrestrial ecosystem?

○ wolves

○ blackbirds

○ mice

○ flowers

37 The mixing of nutrients from the ocean floor to the surface of the water is known as

○ ocean convection.

○ sunlight stirring.

○ turning over.

○ upwelling.

38 **Diseases can be communicable or noncommunicable based on whether or not you can catch them from another person.**

Which of the following is considered communicable?

○ heart disease

○ AIDS

○ diabetes

○ stroke

39 Look at the drawing of the water cycle below.

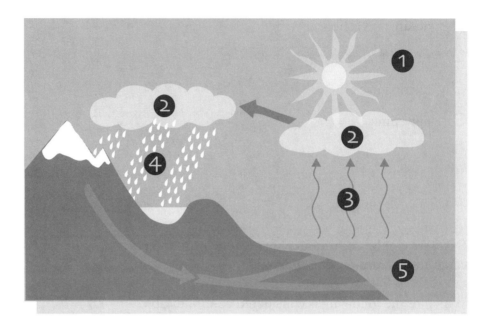

Which number indicates precipitation in this drawing?

○ 2

○ 3

○ 4

○ 5

40 Which of the following could be indicated by a noticeable change in humidity?

○ high pressure

○ low pressure

○ passing of a front

○ an inversion

Go On

41 On the lines below the weather map symbols, label the fronts shown.

_____   _____

42 Which organelle is found in plant cells and **not** in animal cells?

○ nucleus
○ cytoplasm
○ mitochondrion
○ chloroplast

43 Which of the following is an optical device used to study cells?

○ telescope
○ microscope
○ stethoscope
○ orthoscope

Go On

44 Which process heats the soil and rocks on the surface of Earth?

- ○ conduction
- ○ convection
- ○ friction
- ○ radiation

45 Which of the following is necessary for photosynthesis to occur?

- ○ sugar
- ○ digestion
- ○ chlorophyll
- ○ pollinators

46 The table below shows the relationship between the distance and time of the flight of a ball.

Time (seconds)	Distance
0	0
1	2 m
2	4 m
3	6 m
4	7 m
5	8 m
6	9 m
7	10 m
8	11 m
9	12 m
10	13 m

Using the grid below, construct a **line graph** showing the relationship between the distance a ball travels and the time it takes to travel that distance. **Be sure to title your graph, label each axis, and indicate the appropriate units for each axis.**

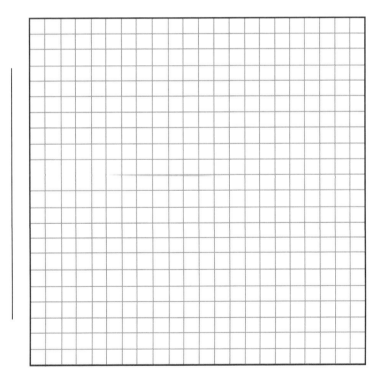

Go On

47 The drawing below is of cars on a roller coaster.

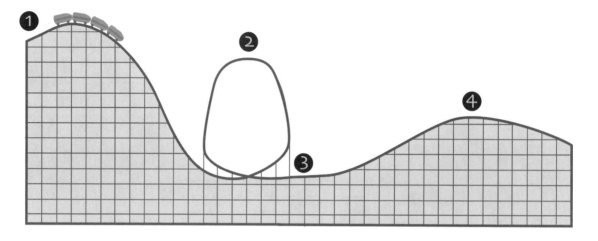

At which point on the roller coaster's path will its kinetic energy be the **greatest**?

○ Point 1

○ Point 2

○ Point 3

○ Point 4

48 When light goes through a prism or a crystal, colors with shorter wavelengths are bent more than colors with longer wavelengths.

Which color is bent the **least**?

○ red

○ green

○ yellow

○ violet

49 When two or more atoms combine chemically to form a new substance, this substance is known as

- ○ an atom.
- ○ an element.
- ○ a mixture.
- ○ a compound.

50 Many times, the first explanation of an observation is not the correct one. Ancient cultures observed the rising and setting of the sun and the motions of the stars and came up with different explanations for these motions.

What are **two** possible explanations for the rising and setting of the sun?

_____

_____

_____

_____

_____

51 Over the past several hundred years, data about the average temperature of the Earth system has been accumulating as scientists do their work. Scientists mostly agree that Earth's atmosphere is warming although the reason why is more complicated.

Which of the following is a reasonable explanation for global warming?

○ the accumulation of greenhouse gases in the atmosphere

○ normally observed cycles of warming and cooling of the Earth system

○ pollution from the burning of fossil fuels

○ All of the above are reasonable explanations.

Directions

The data table below shows the percent of sunlight that is reflected by different surfaces. Use the information shown to do Numbers 52 and 53.

Percent of Sunlight Reflected by Different Surfaces

Surface	Percent Reflected
Clouds	70%
Concrete	20%
Green Crops	15%
Forest	8%
Meadow	14%
Plowed Field	15%
Highway	7%
Sand	50%
Snow	85%
Water	8%

Go On

52 Write **two** conclusions that can be made from the data in the table.

Conclusion 1

_____

_____

Conclusion 2

_____

_____

53 Which statement is **true** based on the data table?

○ Different kinds of water reflect the most light.

○ Green crops reflect the least light.

○ Dirt and sand reflect about the same.

○ Snow is a very important surface that keeps our Earth cool.

54 A section of the periodic table of the chemical elements is shown below.

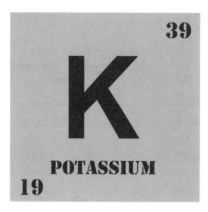

In this case, the number 19 refers to the atomic number for potassium. This is the

○ number of neutrons in the nucleus of potassium.

○ number of compounds it can form.

○ number of protons in the nucleus of potassium.

○ atomic mass of potassium.

Science Assessment Two—Session Three

55 **Vertebrates are classified by their physical characteristics.**

Which of the following is **not** a characteristic of birds?

○ feathers

○ teeth

○ hollow bones

○ lay eggs

56 One of the most active faults in North America is the San Andreas fault in California. This is a slip fault that has moved a very long way in the last 100 years and causes earthquakes on a very regular basis, although they are not very predictable.

What are **two** reasonable questions that could be helpful to people who live near the San Andreas fault?

_____

_____

_____

_____

_____

_____

57 The theory of plate tectonics has helped people figure out other useful information.

Which of the following is **not** directly related to this idea?

- ○ understanding how earthquakes happen
- ○ understanding how volcanoes are formed
- ○ understanding how mountains are formed
- ○ understanding how the atmosphere is changing

Go On

58 Which of the following could **not** be evidence that supports a hypothesis?

○ measurements
○ observations
○ opinions of scientists
○ models

59 Look at the diagram of Earth's atmosphere below.

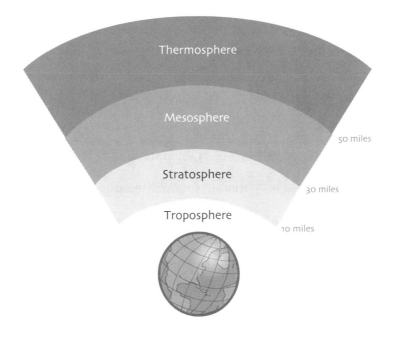

In which layer do most airplanes fly?

○ thermosphere
○ mesosphere
○ stratosphere
○ troposphere

60 How are lungs and gills the same, and how are they different?

_____

_____

_____

_____

_____

_____

61 Which statement about sugar water is **correct**?

○ It is a compound because it tastes different than regular water.

○ It is a mixture because you can separate the sugar and water by evaporating the water.

○ It is a compound because it has more than one substance combined.

○ It is a mixture because a chemical change takes place when it is combined.

Directions

The data table below shows the top speeds of various land mammals. Use the information shown to do Numbers 62 and 63.

Mammal	Top Speed (mph)
Cheetah	65
Pronghorn	55
Lion	50
Gazelle	47
Elk	45
Horse	43
Coyote	43
Greyhound (dog)	42
Mule Deer	35
Human	28
Squirrel	12
Pig	11
Mouse	8
Sloth	0.45

62 Make **two** predictions based on the information from this data table.

Prediction 1:

___

___

___

Prediction 2:

___

___

___

Go On

63 Considering the data table and what you know about animals, which prediction about mammals do you think would be true?

○ Faster mammals are heavier.

○ Slower mammals are endangered.

○ Faster mammals have longer legs.

○ Slower mammals are pets.

64 **Examine the diagram of rock layers below.**

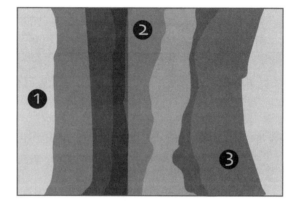

Which layer is the oldest?

○ 1

○ 2

○ 3

○ You can't tell from looking at this profile.

65 In the human body, which systems work together to transport nutrients to the cells of the body?

○ nervous, immune, and respiratory

○ respiratory, digestive, and circulatory

○ circulatory, integumentary, and immune

○ endocrine, skeletal, and excretory

66 **An investigation resulted in data that showed no particular pattern.**

What would be the **best** way to display this data?

○ a line graph

○ a bar graph

○ a circle graph

○ the original data table

67 **When a spacecraft leaves Earth, it no longer needs a push from a rocket engine to travel to its destination.**

Why is this?

○ All objects travel in a straight line at a constant speed until something else changes their motion.

○ Newton's first law of motion states that it will.

○ Since space is mostly empty, nothing is there to change the spacecraft's motion.

○ All of the above are correct.

Go On

68 A class is testing different squirt guns. Each different squirt gun is loaded with water from the same faucet and shot three times. The distance each student shot the squirt gun was measured in meters and then averaged.

What is the dependent variable in this experiment?

- ○ squirt gun
- ○ amount of water
- ○ average distance
- ○ color of the squirt gun

69 Below are three boxes. In the first box, draw dots indicating particles in a solid. In the second box, draw dots to indicate the spacing and energy of particles in a liquid. In the third box, draw dots to indicate the spacing and energy of particle in a gas.

Directions

The data table below shows the distance traveled by an average honeybee per five minutes time. Use the information shown on the table to do Numbers 70 and 71.

Time (minutes)	Distance (km)
5	2.01 km
10	4.02 km
15	6.03 km
20	8.04 km
25	10.05 km
30	12.06 km
35	14.07 km
40	16.08 km
45	18.09 km
50	20.1 km
55	22.11 km
60	24.12 km

70 On the grid below, construct a **line graph** that shows the relationship between Time and Distance. **Be sure to title your graph, label each axis, and indicate the appropriate units for each axis.**

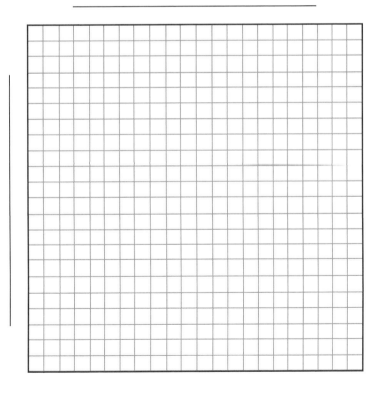

71 Based on the graph you constructed in Question 70, what do you predict will happen after a bee has flown for 120 minutes.

○ It will have flown 26.13 km.

○ It will have flown double the distance it had flown in 60 minutes.

○ It will have slowed down considerably, not keeping the average distance shown in the graph.

○ It will have stopped flying altogether.

72 Which particle is found in the nucleus of an atom and has no charge?

○ electron

○ neutron

○ prion

○ proton

73 The following three things were placed in a beaker—mercury (a liquid with a density of 13.6 g/mL), a diamond (a solid with a density of 3.52 g/mL), and a piece of oak (a solid with a density of 0.67 g/mL).

What happened after all three were placed in the beaker?

○ The oak floated and the diamond sank in the mercury.

○ The oak sank and the diamond floated in the mercury.

○ Both the oak and the diamond floated in the mercury.

○ Both the oak and the diamond sank in the mercury.

Go On

74 When continental plates collide, it places stresses throughout the plate.

Which is an example of a large geologic feature far from the edge of a plate?

○   the Rio Grande rift zone

○   the Cascade Mountains in Washington

○   the San Andreas fault in California

○   the volcanoes in Indonesia

75 The Earth is always changing.

What are **two** ways that the convection in the mantle of Earth changes the surface of Earth?

_____

_____

_____

_____

_____

76 **Digestion is the process of breaking down food.**

What are **two** kinds of digestion that can occur in the body so cells can get nutrients from food?

_____

_____

_____

77 **An experiment is being performed to test the effect of the amount of food on the weight gain of certain monkeys.**

What is the independent variable?

_____

_____

What is the dependent variable?

_____

_____

78 One way of organizing how we understand the world is by looking at different levels of organization.

What level within an organism is smaller than a cell?

- ○ molecule
- ○ biosphere
- ○ organ
- ○ organism

79 When a substance undergoes a physical change, like tearing paper or breaking a glass, the mass is the same before and after the event.

This is an example of

- ○ the law of physical changes.
- ○ the theory of relativity.
- ○ the law of conservation of mass.
- ○ the theory of chemistry.

80 Which of the following is a renewable source of energy?

- ○ natural gas
- ○ oil
- ○ coal
- ○ trees

81 Which organ system is responsible for finding and getting rid of foreign invaders?

○ digestive
○ nervous
○ endocrine
○ immune

82 Which of the following is an example of a chemical change?

○ burning a piece of paper
○ dyeing a piece of paper
○ making pulp with the paper in water
○ punching holes in a piece of paper

83 The smallest unit of a compound that retains the properties of that compound is

○ a proton.
○ a quark.
○ an atom.
○ a molecule.

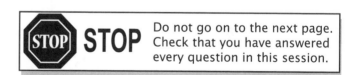

Science Assessment Two—Answer Key

SESSION ONE

Understanding the rock cycle
1 Look at the diagram of the rock cycle below.

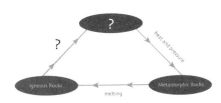

What kind of rock is missing and how is it formed?
Analysis: *Constructed-response answers may vary.* The correct answer is sedimentary rock, which is formed by weathering and deposition.

Explaining that human activity can have positive and negative effects on the natural world
2 Many scientific ideas have had both advantages and disadvantages for the world. One example is nuclear power.

Name **one** advantage and **one** disadvantage for nuclear power.
Analysis: *Constructed-response answers may vary.* Many scientific activities may help the natural world or help to restore negative effects of previous activity. A correct answer will name an advantage of nuclear power, such as it reduces the need for mining coal, and cite one reasonable disadvantage, such as it can also be used for nuclear weapons that can damage ecosystems for thousands of years.

Separating mixtures
3 Which property would be **best** to separate a mixture of iron filings and coal dust?
○ color
○ size
○ solubility
● magnetism

Analysis: *The fourth choice is correct.* Iron filings are magnetic and coal dust is not. The first choice is incorrect. Iron filings and coal dust are the same color. The second choice is incorrect. Iron filings and coal dust are very similar in size. The third choice is incorrect. Neither will dissolve in water.

Explaining that models are used to understand processes and predict changes
4 When may a model have an advantage over direct observations?
○ when processes are very dangerous
○ when processes have only happened once
○ when observations might take too long
● All of the above.

Analysis: *The fourth choice is correct.* There are many situations when a model might have advantages over observing directly from nature. All of these situations provide an advantage to using a model.

Evaluating the potential of an organisms with specific traits to survive and reproduce
5 If an environment changed from very dry to very wet, which organism would survive and reproduce?
● one that was more generally adapted to many different habitats
○ one that was well adapted to the dry environment
○ one that could change its physical characteristics
○ one that reproduces asexually by cloning

Analysis: *The first choice is correct.* More generally adapted organisms still have the genetic variability to have young that survive. The second choice is incorrect. More specifically adapted organisms may have lost the genetic variability to have young that survive. The third choice is incorrect. Organisms cannot change their characteristics once they have developed. The fourth choice is incorrect. Clones do not have the genetic variability they need to survive in changing conditions.

Identifying possible scientific questions
6 A class is thinking about the food in its school's cafeteria.

Which question could be investigated with scientific design in this situation?
○ What is the price of a student lunch?
● How does the amount of fat in the food affect the amount of food sold?
○ Who cooks the food?
○ Where does the food come from?

Analysis: *The second choice is correct.* An investigation can be set up to compare or correlate the amount of fat in a meal and how much food has been sold. The first choice is incorrect. The price of a student lunch is a simple research question. The third choice is incorrect. Who cooks the food is a simple research question. The fourth choice is incorrect. Where the food comes from is a simple research question.

Science Assessment Two—Answer Key

Describing appropriate models for scientific processes

7 Which model matches up with its concept?
- ○ electron cloud ⟶ weather
- ○ greenhouse effect ⟶ plant genetics
- ● plate tectonics ⟶ volcanoes
- ○ energy pyramids ⟶ solar energy

Analysis: *The third choice is correct.* Plate tectonics explain how volcanoes form and predict where they will be. The first choice is incorrect. An electron cloud is a model that explains atoms. The second choice is incorrect. The greenhouse effect is a model that helps explain global warming. The fourth choice is incorrect. Energy pyramids explain energy in ecological systems.

Comparing evidence of past life

8 The picture below is of a fossilized brachiopod from millions of years ago. Fossilized brachiopods are found in limestone all over the world. There are still living brachiopods in the oceans of the world.

To become a fossil, an organism must
- ● fall into the mud and get covered up by more mud.
- ○ live in the ocean.
- ○ have a shell.
- ○ have been around at least 100 million years ago.

Analysis: *The first choice is correct.* Fossils are formed by an organism falling into mud. The second choice is incorrect. Many terrestrial organisms, like dinosaurs, fell into mud and became fossils. The third choice is incorrect. Not all fossilized organisms had shells. The fourth choice is incorrect. There are fossils more recent than 100 million years.

Recognizing and explaining alternative experimental designs

9 A student noticed that she always felt sleepy after lunch. She hypothesized that the starches and sugars in her lunch were making her sleepy. She designed an experiment in which she ate no bread, crackers, or sugars with her lunch for a week and checked how she felt. Then she went back to her old lunch habits and compared how she felt in the two hours after lunch. She decided her hypothesis was supported.

Design another experiment to test her sleepiness.

Analysis: *Constructed-response answers may vary.* The correct answer must include the following: A reasonable explanation for the observation—a hypothesis; the independent variable—something that is manipulated by the researcher; the dependent variable—the effect of the test; the controlled variables—everything else is kept the same. For example, Hypothesis: She is sleepy because she doesn't get enough sleep at night. Experiment: Sleep 10 hours each night for a week and check her reaction. Then, get back to the previous sleeping pattern and compare. Keep everything else the same.

Recognizing and explaining alternative experimental designs

10 An investigator is observing animals in their natural habitat. He hypothesizes that they act very differently when he is there than when he is not there.

Which experimental design will **not** work to test his hypothesis?
- ○ Set up a video camera in a tree to observe then compare with how the animals act to how they acted when he was present.
- ● Sit quietly and do not take notes then compare with how the animals act to how they acted when he was present taking notes.
- ○ Make a blind and hide from the animals then compare with how the animals act to how they acted when he was present and not hidden.
- ○ Perform his observations from a mile away using a powerful telescope then compare with how the animals act to how they acted when he was present.

Analysis: *The second choice is correct.* This design will not work since the investigator is still present. The first choice is incorrect. Setting up a video camera for observation is a reasonable design. The third choice is incorrect. Making a blind and hiding is a reasonable design. The fourth choice is incorrect. Using a telescope for observation is a reasonable design.

Recognizing multicultural contributions to science

11 Which of the following is a cooperative study between cultures?
- ○ the space station
- ○ studies of global warming
- ○ automotive engineering
- ● All of the above are multicultural efforts.

Analysis: *The fourth choice is correct.* Almost all of science is international now.

Science Assessment Two—Answer Key

Describing how organisms can change in response to the environment

12 When changes in the environment occur, the organisms who live there must respond.

Which of the following is a **likely** response to pollution in an environment?
○ Organisms breed more often.
● Many organisms will die.
○ Organisms change their eating habits.
○ Organisms clean their habitat.

Analysis: *The second choice is correct.* Often organisms can't respond and don't survive when their habitat is polluted. The first choice is incorrect. Breeding more often would result in a catastrophe since there is a polluted living area. The third choice is incorrect. Organisms just can't change their eating habits fast enough to survive. The fourth choice is incorrect. Cleaning their habitat is beyond the ability of the organisms.

Planning and designing scientific investigations

13 A scientist thinks that a new drug treatment that he has invented will cure a disease. He wants to test this idea.

Design an experiment he can perform to test his idea.

Analysis: *Constructed-response answers may vary.* The correct answer must include the following: the independent variable—the drug treatment; the dependent variable—improvement in the disease, cure; the controlled variables—what people think, same nutrition, same attention from doctor; design features—how the experiment is done. For example: the scientist should recruit people with the disease he wants to cure. They should all be about as sick as one another. He should then divide them into two groups and give all of them a little white pill. In one group, the pill has the new drug, and in the other group, it doesn't. He should treat all of the people the same in every way and after a while, check their condition to see how they are doing and compare.

Describing why scientific knowledge changes over time

14 Which scientific idea has stayed the same over time?
○ the atomic theory
○ the theory of natural selection
○ germ theory
● None of the above.

Analysis: *The fourth choice is correct.* All scientific ideas tend to develop over time. None of these theories have stayed the same.

Describing environmental factors that limit population size

15 Sometimes changes in an environment affect the number of organisms in a population.

How could the amount of food and nesting sites limit the size of a population?

Analysis: *Constructed-response answers may vary.* The amount of food will support a limited number of organisms. The number of nesting sites will limit the number of offspring that can be raised.

Describing the flow of energy through a circuit

16 In a house circuit, what direction does the electrical energy flow?
○ Electrical energy flows from – to +.
○ Electrical energy flows from + to –.
● Electrical energy alternates directions.
○ None of the above.

Analysis: *The third choice is correct.* AC means alternating current—alternating directions. The first and second choices are incorrect. House circuits are AC—alternating current. The fourth choice is incorrect, since the third choice is correct.

Using technologies

17 Name **two** types of technologies that help people study the ocean.

Analysis: *Constructed-response answers may vary.* The correct answer must include the following: two reasonable technologies that help people study the ocean, including computers, seismographs, compass, GPS, sonar, radar, submarines, robots, etc.

Science Assessment Two—Answer Key

Explaining the power of repeated experiments

18 When an experiment is completed, what is the **most** important thing an investigator can do to be sure of his or her results?
○ publish the results
● repeat the experiment
○ change the variables and start over
○ remove the controls and repeat the experiment

Analysis: *The second choice is correct.* Repeating the experiment will develop comparability and confidence in his or her results. The first choice is incorrect. Publishing gains critique which may be helpful. The third choice is incorrect. There will be no comparability if the variables are changed. The fourth choice is incorrect. There will be no comparability the controls are changed.

Inferring the traits of offspring based on the genes of the parents

19 Consider the Punnett square below dealing with round seeds (R) and wrinkled seeds (r).

	R	r
R	RR	Rr
r	Rr	rr

What percentage of the offspring will be round?
○ 25%
○ 50%
● 75%
○ 100%

Analysis: *The third choice is correct.* In a case like this, round is dominant to the wrinkled seeds. Rr and RR are both round, so 75% are round. The first choice is incorrect. Twenty-five percent of the offspring is wrinkled seeds. The second choice is incorrect. Nothing in the square is 50%. The fourth choice is incorrect. Nothing in this Punnet square is 100%.

Identifying controls in an experiment

20 In an experiment testing dog food, a student made sure he held all of the conditions the same except the kind of food he served the dog.

What are **two** likely control variables in this investigation?
Analysis: *Constructed-response answers may vary.* Both control variables will be things that stay the same between experimental trials or subjects, like same time of day, same dish, same location, etc.

Applying the law of conservation of energy

21 When you lift a pendulum all the way to the top, energy is stored in the ball as potential energy. This energy is converted to motion as the ball falls and is stored again when the ball swings up again.

Why won't the pendulum go on forever?
○ Some of the energy is destroyed by friction.
○ Some of the energy lets you see the pendulum swing.
● Some of the energy is transformed into heat by friction.
○ Some of the energy is used to make the ball heavier at the bottom of the swing.

Analysis: *The third choice is correct.* The more energy that is lost to heat, the less is available to push the ball up again. The first choice is incorrect. Energy cannot be destroyed. The second choice is incorrect. The light energy that lets you see the pendulum is from the surroundings, not the pendulum itself. The fourth choice is incorrect. Matter cannot be created or energy transformed into mass.

Identifying independent and dependent variables

22 A class is testing paper towels to see which one is the strongest when wet. They are using several brands of paper towels. In each case, they wet the towel and then apply weights in 10 gram increments to see when the towel breaks.

What is the independent variable in this experiment?
○ source of the towels
○ weight when the towel breaks
○ amount of water used to wet the towels
● brand of towel

Analysis: *The fourth choice is correct.* The brand of towel is the variable that is being tested by the experimenters. The first choice is incorrect. The source of the towels is not part of the experiment. The second choice is incorrect. The weight when the towel breaks is the dependent variable. The third choice is incorrect. The amount of water is the same for all.

Explaining transferring and transforming of energy

23 In a forest fire, there are many transformations of energy.

What are **two** energy changes that would take place during a forest fire?
Analysis: *Constructed-response answers may vary.* When the forest is burning, chemical energy in the wood changes to heat. Things fall, changing potential energy of position to kinetic energy of motion. Some energy is converted to heat and some into light.

Science Assessment Two—Answer Key

Identifying types of circuits
24 In the electrical system of a house, when a light bulb goes out, the others will still work.
  What kind of circuit must this be?
○ a DC circuit
● a parallel circuit
○ a series circuit
○ an open circuit

Analysis: *The second choice is correct.* A circuit in which there are multiple paths for a complete circuit is a parallel circuit. In the parallel circuit, one appliance can go out and the others are still working. The first choice is incorrect. A house circuit operates on alternating circuits (AC). The third choice is incorrect. If one light went out in a series circuit they all would. The fourth choice is incorrect. An open circuit is one that is broken and the electricity cannot flow.

Relating the numbers of chromosomes to mitosis and meiosis
25 In human cells, a diploid cell has 46 chromosomes. After dividing by the process of mitosis, how many chromosomes are in the resulting cells?
○ 12
○ 23
● 46
○ 92

Analysis: *The third choice is correct.* Mitosis is a body process that starts and ends with diploid cells. Meiosis results in haploid cells. Forty-six is the diploid number. The first choice is incorrect. Twelve is not quite half of half of 46. The second choice is incorrect. Twenty-three is half of 46—the haploid number. The fourth choice is incorrect. Ninety-two would be double diploid.

Understanding technologies needed for the exploration of space
26 Which tool can be used to determine the composition of a star?
○ a telescope
● a spectroscope
○ a microscope
○ an electronic scale

Analysis: *The second choice is correct.* Different elements give off different spectra of light that can be measured using a spectroscope. The first choice is incorrect. A telescope is good for watching or photographing stars. The third choice is incorrect. A microscope is for seeing small things. The fourth choice is incorrect. An electronic scale is for measuring the mass of small quantities.

Recording and reporting data
27 What is the mass of the ball on the scale shown below?

● 65 grams
○ 65 pounds
○ 65 kilograms
○ 6.5 grams

Analysis: *The first choice is correct.* 50 + 10 + 5 grams = 65 grams. The second choice is incorrect. Masses are in grams. The third choice is incorrect. Masses are in grams. The fourth choice is incorrect. A mass of 6.5 grams is an incorrect total.

SESSION TWO

Differentiating between mitosis and meiosis
28 In mitosis,
○ two cells that are genetically different can result.
● two identical cells result.
○ sex cells are produced.
○ half of the cells automatically die.

Analysis: *The second choice is correct.* Mitosis is a body process that reproduces identical cells to the parent. The first choice is incorrect. The cell is exactly duplicated. The third choice is incorrect. Producing sex cells is the function of meiosis. The fourth choice is incorrect. Half of the cells do not automatically die.

Explaining mass and weight
29 Weight and mass are different.
  What is being measured when you measure weight?
○ number of atoms
○ amount of particles
○ force of planetary spin on an object
● force of gravity on an object

Analysis: *The fourth choice is correct.* The force of gravity on an object is what you are measuring when you measure weight. The first choice is incorrect. The number of atoms affects the mass, not necessarily the weight. The second choice is incorrect. The amount of particles affects the mass, not necessarily the weight. The third choice is incorrect. The momentum of spin on an object is not the weight.

Science Assessment Two—Answer Key

Inferring numbers of organisms and energy in an energy pyramid
30 The sketch below is a very different version of a food pyramid.

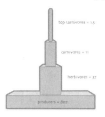

What is being measured in this food pyramid?
- ○ numbers of organisms
- ○ weight of each organism
- ● biomass in the system
- ○ sunlight received by each level

Analysis: *The third choice is correct.* In a food pyramid, the size of the levels shows the amount of biomass at that level. The first choice is incorrect. The number refers to amount of biomass. The second choice is incorrect. The weight of the organism is unimportant. The fourth choice is incorrect. Sunlight is not displayed in a food pyramid.

Describing the components of the universe
31 There are many different parts of the universe that are interesting to astronomers who wonder about its origin.

Which of the following is an area where stars develop?
- ○ a blue giant star
- ○ a solar system
- ● a nebula
- ○ a galaxy

Analysis: *The third choice is correct.* A nebula is a cloud of interstellar dust. New stars frequently begin in nebulae. The first choice is incorrect. A blue giant is just one star. The second choice is incorrect. A solar system is usually a star and its planets. The fourth choice is incorrect. A galaxy is a collection of stars spinning around a central point.

Understanding the phases of the moon and tides
32 Which part of the diagram below shows the moon when its phase is new moon?

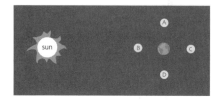

- ○ A
- ● B
- ○ C
- ○ D

Analysis: *The second choice is correct.* Position B is a new moon. The first choice is incorrect. The moon is about half full in position A. The third choice is incorrect. The moon is full in position C. The fourth choice is incorrect. The moon is about half full in position D.

Identifying possible scientific questions
33 A gardener would like to know what to plant on the west side of her greenhouse.

Write a valid scientific question that could be asked in this situation and suggest a way the gardener might investigate the question.

Analysis: *Constructed-response answers may vary.* The correct answer must include the following: The question should be a reasonable, testable, and repeatable; the investigation should reasonably investigate the question. Sample Answer: Question: Which of four different plants grows best on the west side of the greenhouse? Investigation: The gardener could plant a few of each of the four different plants on the west side of the greenhouse and observe them over time.

Describing the colors of white light
34 White light comes from the sun as a spectrum of different wavelengths (colors). It is affected by the atmosphere as it passes through.

Which is an effect of this interaction between sunlight and the atmosphere?
- ○ red sunrises
- ○ orange sunsets
- ○ blue sky
- ● All of the above.

Analysis: *The fourth choice is correct.* All of the choices are effects of light scattering through the atmosphere. Red sunrises are caused by the scattering of other wavelengths, so only the long red waves get through. Orange sunsets are caused by the scattering of other wavelengths, so only the longer orange waves get through. Blue light gets scattered around as the light passes through the atmosphere so we see it more.

Describing the solar system
35 Which planet in our solar system is the smallest?
- ● Mercury
- ○ Venus
- ○ Uranus
- ○ Earth

Analysis: *The first choice is correct.* Mercury is the smallest planet. The planets from smallest to largest are: Mercury, Mars, Venus, Earth, Neptune, Uranus, Saturn, and Jupiter. The second choice is incorrect. Venus is the third smallest planet. The third choice is incorrect. Uranus is the third largest planet. The fourth choice is incorrect. Earth is the fourth smallest planet.

Science Assessment Two—Answer Key

Examining the flow of energy in an ecosystem
36 Which of the following is a producer in a terrestrial ecosystem?
- ○ wolves
- ○ blackbirds
- ○ mice
- ● flowers

Analysis: *The fourth choice is correct.* Flowers are plants and, therefore, producers. The first choice is incorrect. Wolves eat other animals and are consumers. The second choice is incorrect. Blackbirds are consumers; they eat seeds and insects. The third choice is incorrect. Mice are consumers; they eat plants and seeds.

Understanding the composition and characteristics of oceans
37 The mixing of nutrients from the ocean floor to the surface of the water is known as
- ○ ocean convection.
- ○ sunlight stirring.
- ○ turning over.
- ● upwelling.

Analysis: *The fourth choice is correct.* Nutrients are frequently stirred up from the bottom of the ocean keeping the ocean a rich stew of life. The first choice is incorrect. Convection may drive the mixing of nutrients, but this isn't the term. The second choice is incorrect. Sunlight stirring is not the correct term. The third choice is incorrect. Turning over is not the correct term.

Classifying diseases as communicable and noncommunicable
38 Diseases can be communicable or noncommunicable based on whether or not you can catch them from another person.
  Which of the following is considered communicable?
- ○ heart disease
- ● AIDS
- ○ diabetes
- ○ stroke

Analysis: *The second choice is correct.* Communicable means you can catch the disease from another person. AIDS is passed from person to person. The first choice is incorrect. Heart disease is not passed from person to person. The third choice is incorrect. Diabetes is not passed among people. The fourth choice is incorrect. A stroke is not passed from person to person.

Explaining the water cycle
39 Look at the drawing of the water cycle below.

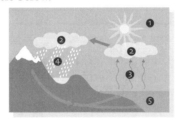

Which number indicates precipitation in this drawing?
- ○ 2
- ○ 3
- ● 4
- ○ 5

Analysis: *The third choice is correct.* Number 4 is precipitation. The first choice is incorrect. Number 2 is condensation in the clouds. The second choice is incorrect. Number 3 is evaporation. The fourth choice is incorrect. Number 5 shows liquid water in the ocean.

Identifying causes of changes in the weather
40 Which of the following could be indicated by a noticeable change in humidity?
- ○ high pressure
- ○ low pressure
- ● passing of a front
- ○ an inversion

Analysis: *The third choice is correct.* Changes in humidity usually mean that a front has just passed. The first choice is incorrect. The pressure cell could change the humidity level if it was moving through with a front. The second choice is incorrect. The pressure cell could change the humidity level if it was moving through with a front. The fourth choice is incorrect. This air is stable.

Interpreting weather data
41 On the lines below the weather map symbols, label the fronts shown.

Analysis: The first drawing is a cold front and the second drawing is a warm front.

Differentiating between animal and plant cells
42 Which organelle is found in plant cells and **not** in animal cells?
- ○ nucleus
- ○ cytoplasm
- ○ mitochondrion
- ● chloroplast

Analysis: *The fourth choice is correct.* Plant cells perform photosynthesis which requires chloroplast. The first choice is incorrect. Both animal and plant cells have a nucleus. The second choice is incorrect. Both animal and plant cells have cytoplasm. The third choice is incorrect. Both animal and plant cells have a mitochondrion.

Science Assessment Two—Answer Key

Using technologies
43 Which of the following is an optical device used to study cells?
○ telescope
● microscope
○ stethoscope
○ orthoscope

Analysis: *The second choice is correct.* A microscope is used to study cells. The first choice is incorrect. Telescopes are for studying distant objects. The third choice is incorrect. A doctor uses a stethoscope to listen to your lungs and your heart. The fourth choice is incorrect. An orthoscope is used by your eye doctor to look at your cornea.

Explaining the heating of Earth by the sun
44 Which process heats the soil and rocks on the surface of Earth?
● conduction
○ convection
○ friction
○ radiation

Analysis: *The first choice is correct.* Conduction requires contact with the object heating it. The second choice is incorrect. Convection requires a substance to circulate. The third choice is incorrect. Friction is from mechanical contact and motion. The fourth choice is incorrect. This form of heat travels through space.

Describe photosynthesis and cellular respiration
45 Which of the following is necessary for photosynthesis to occur?
○ sugar
○ digestion
● chlorophyll
○ pollinators

Analysis: *The third choice is correct.* Chlorophyll is necessary to make the reaction work. The first choice is incorrect. Sugar is a product of photosynthesis. The second choice is incorrect. Plants do not digest food. The fourth choice is incorrect. Pollinators are important for reproduction, not photosynthesis.

Recognizing different ways to communicate results
46 The table below shows the relationship between the distance and time of the flight of a ball.

Time (seconds)	Distance
0	0
1	2 m
2	4 m
3	6 m
4	7 m
5	8 m
6	9 m
7	10 m
8	11 m
9	12 m
10	13 m

Using the space below, construct a **line graph** showing the relationship between the distance a ball travels and the time it takes to travel that distance. **Be sure to title your graph, label each axis, and indicate the appropriate units for each axis.**

Analysis: *Constructed-response answers may vary.*

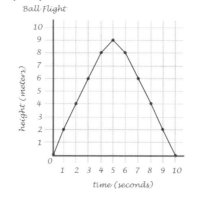

Comparing potential and kinetic energy
47 The drawing below is of cars on a roller coaster.

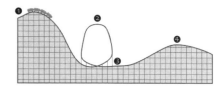

At which point on the roller coaster's path will its kinetic energy be the **greatest**?
○ Point 1
○ Point 2
● Point 3
○ Point 4

Analysis: *The third choice is correct.* Point 3 is lowest here. If the cars started here, they would probably not move. They probably have the highest kinetic energy here if they started at Point 1 and converted all of that potential energy to kinetic. The first choice is incorrect. The cars are the highest here. In a case like this one, potential energy will be the greatest when the cars, which are pulled by gravity, are at their highest point. It will be the least when they are at their lowest point. The second choice is incorrect. Point 3 is lower than Point 2. The fourth choice is incorrect. Point 3 is lower than Point 4.

Science Assessment Two—Answer Key

Comparing the wavelengths of colors of light

48 When light goes through a prism or a crystal, colors with shorter wavelengths are bent more than colors with longer wavelengths.

Which color is bent the **least**?
- ● red
- ○ green
- ○ yellow
- ○ violet

Analysis: *The first choice is correct.* The colors of the spectrum from longest to shortest are: red, orange, yellow, green, blue, indigo, and violet. Red has the longest wavelength of visible light. The second choice is incorrect. Green has a shorter wavelength than red. The third choice is incorrect. Yellow has a shorter wavelength than red. The fourth choice is incorrect. Violet has the shortest wavelength of light.

Explaining ratios of atoms in compounds

49 When two or more atoms combine chemically to form a new substance, this substance is known as
- ○ an atom.
- ○ an element.
- ○ a mixture.
- ● a compound.

Analysis: *The fourth choice is correct.* When two or more atoms combine to form a new substance, it is a compound. The first choice is incorrect. When atoms combine, they form compounds. The second choice is incorrect. When atoms form a different substance, it is no longer an element. The third choice is incorrect. In a mixture, nothing has combined chemically.

Describing other reasonable explanations

50 Many times, the first explanation of an observation is not the correct one. Ancient cultures observed the rising and setting of the sun and the motions of the stars, and came up with different explanations for these motions.

What are **two** possible explanations for the rising and setting of the sun?

Analysis: *Constructed-response answers may vary.* Of course we know that the rising and setting of the sun is due to the spin of Earth on its axis. This is very hard to observe directly and historically, people have had many different explanations which we now know are wrong. The correct answer should include two explanations that would explain the motion of the sun across the sky. Sample Answer: One explanation for the rising and setting of the sun is that Earth is spinning on its axis making the sun appear to move across the sky. A second explanation would be that ancient people thought the sun orbited Earth, and Earth was stationary.

Describing other reasonable explanations

51 Over the past several hundred years, data about the average temperature of the Earth system has been accumulating as scientists do their work. Scientists mostly agree that Earth's atmosphere is warming although the reason why is more complicated.

Which of the following is a reasonable explanation for global warming?
- ○ the accumulation of greenhouse gases in the atmosphere
- ○ normally observed cycles of warming and cooling of the Earth system
- ○ pollution from the burning of fossil fuels
- ● All of the above are reasonable explanations.

Analysis: *The fourth choice is correct.* All three of the explanations are reasonable. The first choice is incorrect. The accumulation of greenhouse gases is a reasonable explanation but not the only one. The second choice is incorrect. Cycles of warming and cooling of the Earth system is a reasonable explanation but not the only one. The third choice is incorrect. Pollution from the burning of fossil fuels is a reasonable explanation but not the only one.

Science Assessment Two—Answer Key

Directions: The data table below shows the percent of sunlight that is reflected by different surfaces. Use the information shown to do Numbers 52 and 53.

Percent of Sunlight Reflected by Different Surfaces

Surface	Percent Reflected
Clouds	70%
Concrete	20%
Green Crops	15%
Forest	8%
Meadow	14%
Plowed Field	15%
Highway	7%
Sand	50%
Snow	85%
Water	8%

Interpreting and evaluating data
52 Write **two** conclusions that can be made from the data in the table.
Analysis: *Constructed-response answers may vary.* The correct answer must include two reasonable conclusions supported by this data, such as: the darker the surface, the less light is reflected, or plants reflect very little sunlight.

Interpreting and evaluating data
53 Which statement is **true** based on the data table?
○ Different kinds of water reflect the most light.
○ Green crops reflect the least light.
○ Dirt and sand reflect about the same.
● Snow is a very important surface that keeps our Earth cool.
Analysis: *The fourth choice is correct.* Snow, which covers a large part of the globe, reflects the most sunlight of all (85%). It will not reflect much at all if it melts into water (8%). The first choice is incorrect. Water reflects one of tghe lowest percentages of light (8%). The second choice is incorrect. Green crops reflect more than water or highways. The third choice is incorrect. Sand reflects about 50%, while dirt in a plowed field reflects about 15%.

Explaining the number of protons
54 A section of the periodic table of the chemical elements is shown below.

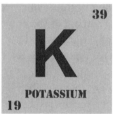

In this case, the number 19 refers to the atomic number for potassium. This is the
○ number of neutrons in the nucleus of potassium.
○ number of compounds it can form.
● number of protons in the nucleus of potassium.
○ atomic mass of potassium.
Analysis: *The third choice is correct.* The atomic number is the number of protons in the nucleus of an atom. The first choice is incorrect. The atomic number is not neutrons. The second choice is incorrect. The number of compounds an element can form is not in the periodic table of elements. The fourth choice is incorrect. The other number, 39, is the mass. It is the total number of protons and neutrons in a typical nucleus of an atom.

SESSION THREE

Identifying the physical characteristics of vertebrates
55 Vertebrates are classified by their physical characteristics.
Which of the following is **not** a characteristic of birds?
○ feathers
● teeth
○ hollow bones
○ lay eggs
Analysis: *The second choice is correct.* Birds do not have teeth. The first choice is incorrect. Birds have feathers. The third choice is incorrect. Birds have hollow bones. The fourth choice is incorrect. Birds lay eggs.

Evaluating data and explaining patterns seen in the past
56 One of the most active faults in North America is the San Andreas fault in California. This is a slip fault that has moved a very long way in the last 100 years and causes earthquakes on a very regular basis, although they are not very predictable.

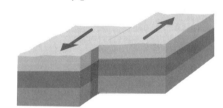

What are **two** reasonable questions that could be helpful to people who live near the San Andreas fault?
Analysis: *Constructed-response answers may vary.* Reasonable questions include: When was the last earthquake? Where is the fault exactly? How much does the fault generally move each time?

Science Assessment Two—Answer Key

Evaluating data and explaining patterns seen in the past

57 The theory of plate tectonics has helped people figure out other useful information.

Which of the following is **not** directly related to this idea?

- ○ understanding how earthquakes happen
- ○ understanding how volcanoes are formed
- ○ understanding how mountains are formed
- ● understanding how the atmosphere is changing.

Analysis: *The fourth choice is correct.* The changing nature of the atmosphere is not related to the drifting of the continents. The first choice is incorrect. Understanding how earthquakes happen is directly related to plate tectonics. The second choice is incorrect. Understanding how volcanoes are formed is directly related to plate tectonics. The third choice is incorrect. Understanding how mountains are formed is directly related to plate tectonics.

Using evidence

58 Which of the following could **not** be evidence that supports a hypothesis?

- ○ measurements
- ○ observations
- ● opinions of scientists
- ○ models

Analysis: *The third choice is correct.* Opinions are not evidence. The first choice is incorrect. Measurements can be used as evidence. The second choice is incorrect. Observations can be used as evidence. The fourth choice is incorrect. Models can be used as evidence.

Identifying layers of the atmosphere

59 Look at the diagram of Earth's atmosphere below.

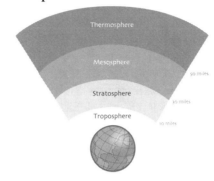

In which layer do most airplanes fly?

- ○ thermosphere
- ○ mesosphere
- ○ stratosphere
- ● troposphere

Analysis: *The fourth choice is correct.* The troposphere is the layer we live in and where most airplanes fly. Planes generally fly below 40,000 feet which is still less than 10 miles above Earth. The first choice is incorrect. Only the space shuttle can fly in the thermosphere. The second choice is incorrect. The mesosphere layer is very cold and too high above Earth. The third choice is incorrect. Maybe a few military planes could fly in the stratosphere, but the majority of planes cannot fly there.

Identifying and comparing gas exchange organs

60 How are lungs and gills the same, and how are they different?

Analysis: Gills and lungs are the same because both are used to exchange oxygen and carbon dioxide. Gills and lungs are different because gills work in the water and lungs work in the air.

Explaining mixtures and compounds

61 Which statement about sugar water is **correct**?

- ○ It is a compound because it tastes different than regular water.
- ● It is a mixture because you can separate the sugar and water by evaporating the water.
- ○ It is a compound because it has more than one substance combined.
- ○ It is a mixture because a chemical change takes place when it is combined.

Analysis: *The second choice is correct.* Sugar and water can be separated using their original physical properties. The first choice is incorrect. The sweet taste is a property of the original sugar. The third choice is incorrect. Compounds and mixtures both are composed of different substances. The difference is that compounds are the result of a chemical change. The fourth choice is incorrect. A chemical change occurs in a compound not a mixture.

Science Assessment Two—Answer Key

Directions: The data table below shows the top speeds of various land mammals. Use the information shown to do Numbers 62 and 63.

Mammal	Top Speed (mph)
Cheetah	65
Pronghorn	55
Lion	50
Gazelle	47
Elk	45
Horse	43
Coyote	43
Greyhound (dog)	42
Mule Deer	35
Human	28
Squirrel	12
Pig	11
Mouse	8
Sloth	0.45

Making predictions
62 Make **two** predictions based on the information from this data table.
Analysis: *Constructed-response answers may vary.* The correct answer may include the following: Prediction 1: Predators are faster than prey. Prediction 2: Smaller animals are slower.

Making predictions
63 Considering the data table and what you know about animals, which prediction about mammals do you think would be **true**?
○ Faster mammals are heavier.
○ Slower mammals are endangered.
● Faster mammals have longer legs.
○ Slower mammals are pets.
Analysis: *The third choice is correct.* The prediction that faster mammals have longer legs is supported by the information in the data table. For example, cheetahs, pronghorns, lions, and gazelles all have longer legs than squirrels, pigs, and mice. The first choice is incorrect. The prediction that faster mammals are heavier is not supported by this information. The second choice is incorrect. The prediction that slower mammals are endangered is not supported by this information. The fourth choice is incorrect. The prediction that slower mammals are pets is not supported by this information.

Interpreting rock layers
64 Examine the diagram of rock layers below.

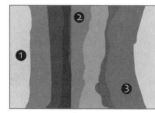

Which layer is the oldest?
○ 1
○ 2
○ 3
● You can't tell from looking at this profile.
Analysis: *The fourth choice is correct.* You cannot tell which layers are on top of the others without more information. According to the concept of superpositioning, younger layers are generally formed on top of older layers. This profile is side by side, not top to bottom.

Identifying and comparing nutrient and waste transport
65 In the human body, which systems work together to transport nutrients to the cells of the body?
○ nervous, immune, and respiratory
● respiratory, digestive, and circulatory
○ circulatory, integumentary, and immune
○ endocrine, skeletal, and excretory
Analysis: *The second choice is correct.* The respiratory, digestive, and circulatory systems work together to transport nutrients to the cells of the body. The first choice is incorrect. The nervous and immune systems are not involved in transporting nutrients and wastes. The third choice is incorrect. The integumentary and immune systems are not involved in transporting nutrients and wastes. The fourth choice is incorrect. The endocrine and skeletal systems are not involved in transporting nutrients and wastes.

Recognizing different ways to communicate results
66 An investigation resulted in data that showed no particular pattern.
What would be the **best** way to display this data?
○ a line graph
○ a bar graph
○ a circle graph
● the original data table
Analysis: *The fourth choice is correct.* The original data table would be more helpful to people looking for meaning in the data. The first choice is incorrect. Line graphs show data that is constantly changing. The second choice is incorrect. Bar graphs compare specific amounts at different times that are not connected. The third choice is incorrect. Circle graphs show percents and portions of a whole thing.

Identifying the forces that change motion
67 When a spacecraft leaves Earth, it no longer needs a push from a rocket engine to travel to its destination.
Why is this?
○ All objects travel in a straight line at a constant speed until something else changes their motion.
○ Newton's first law of motion states that it will.
○ Since space is mostly empty, nothing is there to change the spacecraft's motion.
● All of the above are correct.
Analysis: *The fourth choice is correct.* According to Newton's first law of motion, an object in motion continues in a straight line at a constant speed unless acted upon by an unbalanced force.

CSAP Science for Grade 8   Science Assessment Two—Answer Key

Science Assessment Two—Answer Key

Identifying independent and dependent variables

68 A class is testing different squirt guns. Each different squirt gun is loaded with water from the same faucet and shot three times. The distance each student shot the squirt gun was measured in meters and then averaged.

What is the dependent variable in this experiment?
○ squirt gun
○ amount of water
● average distance
○ color of the squirt gun

Analysis: *The third choice is correct.* The average distance is the result of the changing of the independent variable in the experiment. The first choice is incorrect. The squirt gun is the independent variable. The second choice is incorrect. The amount of water was not measured. The fourth choice is incorrect. The color of the squirt gun does not affect the outcome.

Describing the particulate model for matter

69 Below are three boxes. In the first box, draw dots indicating particles in a solid. In the second box, draw dots to indicate the spacing and energy of particles in a liquid. In the third box, draw dots to indicate the spacing and energy of particles in a gas.

Analysis: *Constructed-response answers may vary.*

Directions: The data table below shows the distance traveled by an average honeybee per five minutes time. Use the information on the table to do Numbers 70 and 71.

Time (minutes)	Distance (km)
5	2.01 km
10	4.02 km
15	6.03 km
20	8.04 km
25	10.05 km
30	12.06 km
35	14.07 km
40	16.08 km
45	18.09 km
50	20.1 km
55	22.11 km
60	24.12 km

Visual methods to summarize data

70 On the grid below, construct a **line graph** that shows the relationship between Time and Distance. **Be sure to title your graph, label each axis, and indicate the appropriate units for each axis.**

Analysis:

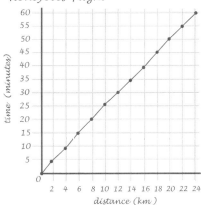

Visual methods to summarize data

71 What do you predict will happen after a bee has flown for 120 minutes?
○ It will have flown 26.13 km
● It will have flown double the distance it had flown 60 minutes.
○ It will have slowed down considerably, not keeping the average distance shown in the graph.
○ It will have stopped flying altogether.

Analysis: *The second choice is correct.* According to the graph constructed in question 70, it can be predicted that a bee will double the distance it had flown in one hour, if it continues to fly into a second hour. The first choice is incorrect. Following the pattern of the table and the graph, a bee will have flown 26.13 km in 65 minutes, not 120 minutes. The third choice is incorrect. There is no evidence to support a prediction that the bee will slow down after flying 120 minutes. The fourth choice is incorrect. There is no evidence to support a prediction that the bee will stop flying after 120 minutes.

Identifying parts of an atom

72 Which particle is found in the nucleus of an atom and has no charge?
○ electron
● neutron
○ prion
○ proton

Analysis: *The second choice is correct.* Neutrons are in the nucleus and carry no charge. The first choice is incorrect. An electron has a negative charge and is outside the nucleus of the atom. The third choice is incorrect. Prions are protein particles that cause diseases like mad cow disease. The fourth choice is incorrect. A proton is the positive particle in the nucleus.

Science Assessment Two—Answer Key

Separating mixtures by density
73 The following three things were placed in a beaker—mercury (a liquid with a density of 13.6 g/mL), a diamond (a solid with a density of 3.52 g/mL), and a piece of oak (a solid with a density of 0.67 g/mL).

What happened after all three were placed in the beaker?
- ○ The oak floated and the diamond sank in the mercury.
- ○ The oak sank and the diamond floated in the mercury.
- ● Both the oak and the diamond floated in the mercury.
- ○ Both the oak and the diamond sank in the mercury.

Analysis: *The third choice is correct.* Both the diamond and the oak have densities less than mercury. The first, second, and third choices are incorrect. Both the diamond and the oak have densities less than mercury.

Understanding plate tectonics
74 When continental plates collide, it places stresses throughout the plate.

Which is an example of a large geologic feature far from the edge of a plate?
- ● the Rio Grande rift zone
- ○ the Cascade Mountains in Washington
- ○ the San Andreas fault in California
- ○ the volcanoes in Indonesia

Analysis: *The first choice is correct.* The Rio Grande rift is in the San Luis Valley in Colorado. It is in the middle of the continent and is caused by shearing forces caused by collisions with other plates. The second choice is incorrect. The Cascade Mountains in Washington are near a plate boundary. The third choice is incorrect. The San Andreas fault in California is near a plate boundary. The fourth choice is incorrect. The volcanoes in Indonesia are near a plate boundary.

Explaining why Earth's surface is always changing
75 Earth is always changing.

What are **two** ways that the convection in the mantle of Earth changes the surface of Earth?

Analysis: *Constructed-response answers may vary.* Correct answers must include two reasonable ways the convection in the mantle of Earth changes the surface. For example, the convection currents drive the tectonic plates, cause volcanoes, and mid-ocean spreading zones.

Describing the processing of food in an organism
76 Digestion is the process of breaking down food.

What are two kinds of digestion that can occur in the body so cells can get nutrients from food?

Analysis: The correct answer must name the following: mechanical digestion and chemical digestion.

Identifying independent and dependent variables
77 An experiment is being performed to test the effect of the amount of food on the weight gain of certain monkeys.

What is the independent variable?
What is the dependent variable?

Analysis: *Constructed-response answers may vary.* The correct answer must include the following: the independent variable (IV) is the variable manipulated by the experimenter or considered as the varying part that affects something else. In this case, the IV is the amount of food. A dependent variable (DV) is the result of the experiment. In this case, the DV is the amount of weight gained by the monkeys.

Identifying the levels of organization
78 One way of organizing how we understand the world is by looking at different levels of organization.

What level within an organism is smaller than a cell?
- ● molecule
- ○ biosphere
- ○ organ
- ○ organism

Analysis: *The first choice is correct.* Molecules are smaller than cells. The second choice is incorrect. The biosphere is larger than a cell. The third choice is incorrect. Organs are larger than a cell. The fourth choice is incorrect. Organisms are larger than a cell.

Applying the law of conservation of mass to physical changes
79 When a substance undergoes a physical change, like tearing paper or breaking a glass, the mass is the same before and after the event.

This is an example of
- ○ the law of physical changes
- ○ the theory of relativity
- ● the law of conservation of mass
- ○ the theory of chemistry

Analysis: *The third choice is correct.* This is an example of the law of conservation of mass. The first choice is incorrect. There is no such thing as the law of physical changes. The second choice is incorrect. The law of relativity has to do with changes in energy and mass. The fourth choice is incorrect. There is no such law as the theory of chemistry.

Science Assessment Two—Answer Key

Understanding the differences between renewable and nonrenewable energy

80 Which of the following is a renewable source of energy?
- ○ natural gas
- ○ oil
- ○ coal
- ● trees

Analysis: *The fourth choice is correct.* A renewable source of energy is one that can be replaced or reused in our lifetime without running out. Trees can be regrown fairly quickly as long as we protect the conditions in which they need to grow. The first choice is incorrect. The supply of natural gas is finite and cannot be replaced. The second choice is incorrect. Once oil is gone, it cannot be renewed. Oil is a nonrenewable resource. The third choice is incorrect. Coal takes millions of years to make.

Identifying organs and organ systems

81 Which organ system is responsible for finding and getting rid of foreign invaders?
- ○ digestive
- ○ nervous
- ○ endocrine
- ● immune

Analysis: *The fourth choice is correct.* The immune system finds and rids the body of foreign invaders. The first choice is incorrect. The digestive system gets nutrients from food. The second choice is incorrect. The nervous system is responsible for sensation and control. The third choice is incorrect. The endocrine system produces chemicals which help to control the body.

Distinguishing between a physical change and a chemical change

82 Which of the following is an example of a chemical change?
- ● burning a piece of paper
- ○ dyeing a piece of paper
- ○ making pulp with the paper in water
- ○ punching holes in a piece of paper

Analysis: *The first choice is correct.* Chemical change involves changing the composition of a substance. In this case, the water and some carbon dioxide is lost from the paper leaving ash. This ash cannot be converted back into paper. The second choice is incorrect. The basic nature of the paper is unchanged and the dye could be removed. The third choice is incorrect. The basic nature of the paper is unchanged and the paper could be reassembled from the pulp. The fourth choice is incorrect. The basic nature of the paper is unchanged and the paper could be restored to its original state without holes.

Identifying that the smallest unit of a compound that still retains the properties of that compound is a molecule

83 The smallest unit of a compound that retains the properties of that compound is
- ○ a proton.
- ○ a quark.
- ○ an atom.
- ● a molecule.

Analysis: *The fourth choice is correct.* A molecule is the smallest unit of a compound that is still that compound. The first choice is incorrect. Protons are the positive charge in the nucleus of an atom. The second choice is incorrect. A quark is one of the two basic constituents of matter (the other is the lepton). Quarks are the theoretical particles that make up protons and neutrons in all atoms. The third choice is incorrect. An atom is the smallest unit of an element.

Science Assessment Two—Session One: Correlation Chart

Use this chart to identify areas for improvement for individual students or for the class as a whole. For example, enter students' names in the left-hand column. When a student misses a question, place an "X" in the corresponding box. A column with a large number of "Xs" shows that the class needs more practice with that particular objective.

Correlation	3.2.a	2.5.a	1.3.b	3.2.b	4.3.a	1.2.a	3.3.a	4.4.a	1.4.a	1.4.a	3.1.a	2.2.a	3.1.b	1.1.b	3.4.a	1.2.b	2.5.b	2.8.b	3.4.b	1.1.a	3.5.a	2.6.a	1.3.a	1.3.a	2.8.a	2.10.a	1.5.a
Question	1	2	3	4	5	6	7	8	9	10	11	12	13	14	15	16	17	18	19	20	21	22	23	24	25	26	27

Science Assessment Two—Session Two: Correlation Chart

Correlation	3.9.a	2.4.a	3.8.b	4.15.a	4.14.a	1.1.c	2.10.a	4.13.a	3.8.a	4.12.a	3.7.a	4.11.a	4.10.a	4.9.a	3.6.b	1.2.b	4.8.a	3.5.a	1.6.a	2.7.c	2.10.b	2.6.a	1.5.a	1.5.a	1.3.a	1.3.a	2.5.c
Question	28	29	30	31	32	33	34	35	36	37	38	39	40	41	42	43	44	45	46	47	48	49	50	51	52	53	54

students' names

Science Assessment Two—Session Three: Correlation Chart

Correlation	3.1.a	1.4.a	1.4.a	1.3.b	4.7.a	3.4.c	2.6.c	1.3.c	1.3.c	4.6.a	3.4.b	1.6.a	2.7.b	1.1.b	2.1.a	1.2.c	1.2.c	2.5.a	2.2.b	4.4.a	4.3.a	3.4.a	1.1.b	3.3.a	2.3.b	4.2.a	3.2.a	2.3.a	2.6.d
Question	55	56	57	58	59	60	61	62	63	64	65	66	67	68	69	70	71	72	73	74	75	76	77	78	79	80	81	82	83

students' names

CSAP Science Standards Checklist

The CSAP Science Standards Checklist can be used by teachers to easily identify what standard each question addresses in the Science Practice Tutorial, Science Assessment One (Session One, Session Two, and Session Three), and Science Assessment Two (Session One, Session Two, and Session Three).

		Question Numbers						
		Practice Tutorial	Assessment One			Assessment Two		
			session one	session two	session three	session one	session two	session three
CSAP Science Standards	1.1.a	1	20			13		
	1.1.b	2	14	34	63	22		68, 77
	1.1.c	3		42	76	6	33	
	1.2.a	4	6			27		
	1.2.b	5	16		79	17	43	
	1.2.c	6			69, 70			70, 71
	1.3.a	7	23, 24				52, 53	
	1.3.b	8	3					58
	1.3.c	9			73, 74			62, 63
	1.4.a	10	9, 10					56, 57
	1.5.a	11	27	28			50, 51	
	1.5.b	12		31, 32		9, 10		
	1.6.a	13		38, 53			46	66
	2.1.a	14		35				69
	2.1.b	15						
	2.2.a	16	12			3		
	2.2.b	17		40				73
	2.3.a	18		44				82
	2.3.b	19						79
	2.3.c	20		29				
	2.4.a	21					29	
	2.4.b	22						
	2.5.a	23	2					72
	2.5.b	24	17					
	2.5.c	25					54	
	2.6.a	26	22				49	

		Question Numbers						
			Assessment One			Assessment Two		
		Practice Tutorial	session one	session two	session three	session one	session two	session three
CSAP Science Standards	2.6.b	27		37				
	2.6.c	28						61
	2.6.d	29			80			83
	2.7.a	30			81			
	2.7.b	31						67
	2.7.c	32			71		47	
	2.8.a	33	25					
	2.8.b	34	18			23		
	2.8.c	35			83			
	2.8.d	36				21		
	2.9.a	37		36		16		
	2.9.b	38		30		24		
	2.10.a	39	26				34	
	2.10.b	40		33			48	
	3.1.a	41	11					55
	3.1.b	42	13					
	3.2.a	43	1					81
	3.2.b	44	4					
	3.3.a	45	7					78
	3.4.a	46	15					76
	3.4.b	47	19					65
	3.4.c	48						60
	3.5.a	49	21				45	
	3.5.b	50		39				
	3.6.a	51		41				
	3.6.b	52					42	
	3.7.a	53		43			38	

Question Numbers

	Practice Tutorial	Assessment One			Assessment Two		
		session one	session two	session three	session one	session two	session three
3.8.a	54		45			36	
3.8.b	55					30	
3.9.a	56		48			28	
3.9.b	57		51		25		
3.10.a	58		54				
3.10.b	59			57	19		
3.11.a	60			61	15		
3.11.b	61			65			
3.11.c	62				12		
3.12.a	63			68	8		
3.13.a	64			72	5		
4.1.a	65			75	1		
4.1.b	66						
4.2.a	67			77			80
4.2.b	68						
4.3.a	69	5					75
4.4.a	70	8					74
4.5.a	71						
4.6.a	72		46				64
4.6.b	73						
4.7.a	74		47				59
4.8.a	75		49			44	
4.9.a	76		50			41	
4.10.a	77		52			40	
4.10.b	78						
4.10.c	79						
4.11.a	80			55		39	

		Question Numbers						
		Practice Tutorial	Assessment One			Assessment Two		
			session one	session two	session three	session one	session two	session three
	4.12.a	81			56		37	
	4.13.a	82			58		35	
	4.13.b	83						
	4.13.c	84						
	4.14.a	85			59		32	
	4.14.b	86						
	4.15.a	87			60		31	
	4.16.a	88			62	26		
	5.1.a	89			64	20		
	5.1.b	90				18		
	5.1.c	91						
	5.2.a	92			66	14		
	5.3.a	93			67	11		
	5.4.a	94						
	5.4.b	95			78	7		
	5.4.c	96				4		
	5.5.a	97			82	2		

NOTES

NOTES

NOTES

Show What You Know® on the CSAP for Grade 8—Additional Products

Student Self-Study Workbook for Mathematics

Student Self-Study Workbook for Reading and Writing

Flash Cards for Reading, Writing, Mathematics, and Science

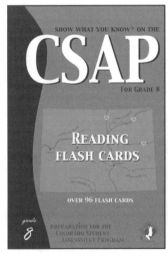
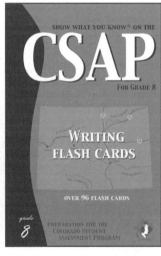
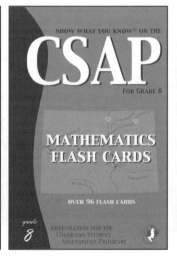
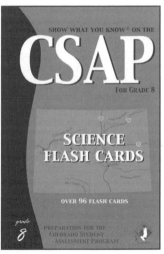

For more information, call our toll free number: 1.877.PASSING (727.7464) or visit our Web site: www.passthecsap.com